# Distributed Localization and Formation Control of Multi-Agent Systems with Local Sensing Information

Yunkai Lv · Huaicheng Yan · Hao Zhang ·
Meng Wang · Zhichen Li

# Distributed Localization and Formation Control of Multi-Agent Systems with Local Sensing Information

Springer

*Author*
See next page

ISBN 978-981-96-6014-8    ISBN 978-981-96-6015-5 (eBook)
https://doi.org/10.1007/978-981-96-6015-5



This Springer imprint is published by the registered company Springer Nature Singapore Pte Ltd.
The registered company address is: 152 Beach Road, #21-01/04 Gateway East, Singapore 189721, Singapore

Yunkai Lv
Key Laboratory of Smart Manufacturing in Energy Chemical Process of the Ministry of Education, School of Information Science and Engineering
East China University of Science and Technology
Shanghai, China

Key Laboratory of System Control and Information Processing of the Ministry of Education
Shanghai Jiao Tong University
Shanghai, China

Hao Zhang
College of Electronic and Information Engineering, Department of Control Science and Engineering, Shanghai Key Laboratory of Wearable Robotics and Human-Machine Interaction
Tongji University
Shanghai, China

Shanghai Institute of Intelligent Science and Technology, National Key Laboratory of Autonomous Intelligent Unmanned Systems and Frontiers Science Center for Intelligent Autonomous Systems
Ministry of Education
Shanghai, China

Zhichen Li
Key Laboratory of Smart Manufacturing in Energy Chemical Process of the Ministry of Education, School of Information Science and Engineering
East China University of Science and Technology
Shanghai, China

Huaicheng Yan
Key Laboratory of Smart Manufacturing in Energy Chemical Process of the Ministry of Education, School of Information Science and Engineering
East China University of Science and Technology
Shanghai, China

Meng Wang
Key Laboratory of Smart Manufacturing in Energy Chemical Process of the Ministry of Education, School of Information Science and Engineering
East China University of Science and Technology
Shanghai, China

*Upon the publication of this book, the author would especially like to thank the China Postdoctoral Science Foundation, the National Natural Science Foundation of China, and the Shanghai Municipal Science and Technology Commission and Education Commission for their long-term support. The author is also grateful to colleagues in the academic and industrial communities at home and abroad, whose valuable exchanges have deepened the author's understanding of networked multi-agent systems and provided inspiration. Special thanks are also extended to the author's team leaders and fellow graduate students for their strong support and assistance.*

# Preface

In an era of rapid technological development, networked intelligent agent systems are reshaping our lives, work, and the operation of society in unprecedented ways. With the continuous breakthroughs and innovations in information technology, the reach of networks extends into various fields, giving rise to the concept of the agent.

Multi-agent systems consist of multiple agents that possess autonomous decision-making, action, and interaction capabilities. These systems exhibit high flexibility, strong reliability, easy maintenance, and low energy consumption, allowing them to collaboratively complete complex tasks through cooperation among the agents. Fusing theories from multiple disciplines, including communication technology, sensor technology, and control science, multi-agent systems play a critical role in complex tasks such as search and rescue, geological exploration, and environmental inspection. They have gradually become a key focal point for achieving China's modernization in digitization, networking, and intelligence.

In 2017, the State Council issued the "New Generation Artificial Intelligence Development Plan," clearly identifying collaborative perception, localization, and control of autonomous unmanned systems as national strategic goals and key tasks. The Ministry of Science and Technology launched the "Science and Technology Innovation 2030 'New Generation Artificial Intelligence' Major Project Guidelines" in 2018, listing "autonomous unmanned systems" as one of the continuous research directions in AI. Furthermore, the Chinese Academy of Sciences' "2019 AI Development White Paper" published in 2020 identified "autonomous unmanned system technology" as a key AI technology. Evidently, collaborative perception and control in multi-autonomous unmanned systems represent a core research area for the future development of unmanned systems.

While there are enormous opportunities for development, the research on multi-agent systems inevitably faces many challenges, such as the system's reliability and stability, as well as adaptability to complex environments. However, these challenges also inspire researchers to explore and innovate continuously, driving the evolution and perfection of networked multi-agent systems. Currently, most of the results in distributed localization for multi-agent systems are focused on static scenarios or systems with slow-moving agents, making it difficult to ensure accurate localization estimates for dynamic scenarios with fast-changing agents. Meanwhile, most of the research on distributed formation control for multi-agent systems assumes that a global reference frame is known, making it challenging to ensure formation control under a local reference frame. Achieving precise localization estimation and formation control for multi-agent systems in complex dynamic scenarios is one of the pressing challenges that need to be addressed.

This book mainly draws from the author's systematic research on key technologies and important issues related to localization estimation and formation control for networked multi-agent systems, carried out since the beginning of the author's Ph.D. studies. Chapter 1 introduces the concepts of multi-agent systems, basic issues in localization and formation, research progress both domestically and internationally, and the theory of system stability. Chapters 2–8 present research results on distributed localization estimation under conditions such as random deployment, non-ideal channels, random noise, varying-trajectory length, sensor multiplicative failure, and local bearing measurement. Chapter 9 focuses on integrated research progress in localization estimation and formation control for multi-agent systems under local perception conditions. The book demonstrates the effectiveness and superiority of the theoretical methods through simulation and physical examples.

This book is mainly intended for scholars in fields such as networked control, intelligent control, multi-agent systems, and unmanned systems, as well as Ph.D. and Master's students. It is also useful for control engineers working in industries such as power and aviation.

Due to the author's limited abilities, there are inevitably some oversights in the book. Readers are kindly requested to provide criticism and suggestions.

Shanghai, China
September 2024

Yunkai Lv

**Acknowledgements** This book has been supported by the Postdoctoral Innovative Talent Support Program (BX20240115), National Natural Science Foundation of China (62333005, 62403200, 62433014), Shanghai Natural Science Foundation(24ZR1416200), the Shanghai International Science and Technology Cooperation Project (22510712000), the China Postdoctoral Science Foundation General Program (2024M750905), and the Fundamental Research Funds for the Central Universities. among others. This book has been supported by the Postdoctoral Innovation Talent Support Program (BX20240115), National Natural Science Foundation of China (62403200, 62333005, 62433014, 62373152, 62173146), Shanghai Natural Science Foundation (24ZR1416200), the Shanghai International Science and Technology Cooperation Project (22510712000), Shanghai Rising-Star Program (23QA1402300), China Postdoctoral Science Foundation General Program (2024M750905), Shanghai Aerospace Science and Technology Innovation Fund (SAST2024-066), China University-Industry-Research Innovation Founds (2023IT041), and the Fundamental Research Funds for the Central Universities.

**Competing Interests** The authors have no conflicts of interest to declare that are relevant to the content of this chapter.

**Ethics Approval** The chapter does not include studies with humans, and no ethics approval is required.

# Contents

# Acronyms

| | |
|---|---|
| $\mathbb{N}$ | Integer field |
| $\mathbb{N}^+$ | Positive integer field |
| $\mathbb{R}$ | Real number field |
| $\mathbb{R}^n$ | $n$-dimensional Euclidean space of real vectors |
| $\mathbb{R}^{n\times n}$ | $n \times n$ dimensional real matrix space |
| $\mathbf{A}^\top$ | Transpose matrix of $\mathbf{A}$ |
| $\lambda_{\max}(\mathbf{A})$ | Maximum eigenvalue of the square matrix $\mathbf{A}$ |
| $\lambda_{\min}(\mathbf{A})$ | Minimum eigenvalue of the square matrix $\mathbf{A}$ |
| $I_n$ or $\mathbf{I}$ | $n \times n$ dimensional identity matrix or identity matrix with suitable dimensions |
| $\mathbf{1}$ | Column vector of appropriate dimensions with all elements equal to 1 |
| $\mathbf{0}$ | Column vector of appropriate dimensions with all elements equal to 0 |
| $\text{diag}_N\{.\}$ | $N$-dimensional block diagonal matrix |
| $\text{diag}\{A_1, A_2, ..., A_N\}$ | Diagonal matrix composed of $A_1, A_2, ..., A_N$ |
| $sign$ | Sign function |
| $\|x\|$ | 2-norm of the vector $x$ |
| $\|x\|_\infty$ | Infinity norm of the vector $x$ |
| $\mathbb{E}(x)$ | Mathematical expectation of the random variable $x$ |
| $prob\{x = 0\}$ | Probability of $x = 0$ |

# Chapter 1
# Introduction

**Abstract** This chapter introduces the concepts, development history, and applications of multi-agent systems. It also elaborates on fundamental issues encountered in networked multi-agent systems, including transmission delays and losses, random arbitrary deployment, measurement and communication noise, random variations in trajectory length, sensor multiplicative failures, and challenges in obtaining global information. The chapter provides an overview of distributed localization estimation and formation control for multi-agent systems and discusses the current research status in these areas. Additionally, some fundamental knowledge about system stability is introduced to lay the foundation for the subsequent research content.

## 1.1 Multi-agent Systems Concept

### *1.1.1 Introduction to Multi-agent Systems*

In the biological world, individuals adapt to the constantly evolving natural environment through their own evolution and collaboration with others to ensure the continuation and growth of their species. The biological world has always been a source of inspiration for human creativity. By observing and reflecting on the advantages and abilities of biological individuals, humans imitate them to create new tools and technologies that align with human survival and development. There exist numerous examples of collective intelligence in the biological world, such as geese migrating in a V-shaped formation, wolves cooperating to hunt prey, ants maintaining a single-file formation to build nests with leaves, and bees collectively gathering nectar [1–4], as shown in Fig. 1.1. Individually, these organisms may have limited capabilities, but through communication and collaboration, they can accomplish complex tasks far beyond their individual abilities.

The autonomy, fault tolerance, local communication, and collaboration emergent from collective behavior have attracted widespread attention from researchers. Studying collective behavior in the biological world can help promote the improvement and development of human social production activities. Hewitt first introduced the concept of agents [5], sparking systematic research on collective intelligence in



Y. Lv et al., *Distributed Localization and Formation Control of Multi-Agent Systems with Local Sensing Information*, https://doi.org/10.1007/978-981-96-6015-5_1

Fig. 1.1 Example of biocluster intelligence

the biological world by experts and scholars in various fields such as sociology and control theory. Gradually, multi-agent systems were used to describe and simulate these phenomena [6–8]. From the late 1980s to the early 1990s, several conferences were held, including the First International Conference on Multi-Agent Systems, the International Conference on Formal Models of Agents, and the International Conference on Agent Theory, Architecture, and Language, marking the rapid development of research on multi-agent systems.

### 1.1.2 Background and Significance of Multi-agent Systems

Multi-agent systems integrate sensors, environmental perception, computers, and automatic control technology, reflecting the latest achievements in information technology development. These systems consist of many individual agents with autonomous planning, action, and interaction capabilities. Through coordination and cooperation, they can accomplish complex tasks. The advantages of multi-agent systems lie in their strong autonomy, robustness, and scalability. In recent years, with the rapid development of sensors, automatic control, and wireless communication technology, multi-agent collaborative decision-making and control have played a crucial role in complex tasks such as search and rescue [9, 10], geological exploration [11, 12], autonomous inspection [13, 14], and logistics transportation [15, 16]. They have been widely applied in scenarios like collaborative combat of unmanned

**Fig. 1.2** Examples of cluster intelligence applications

aerial vehicles (UAVs), satellite reconnaissance and exploration, and pest control in agricultural fields using UAVs, as shown in Fig. 1.2.

Consensus-based collaborative control is a fundamental problem in multi-agent system control [17–20]. Its essence lies in ensuring that the states of all agents converge to a consistent value through information exchange, such as achieving desired positions, velocities, and constant control variables. Localization estimation and formation control are two key research topics in consensus-based collaborative control of multi-agent systems, which have attracted widespread attention from academia and industry. Depending on information processing and storage methods, existing localization and formation control algorithms can be classified into centralized and distributed methods [21–26]. Distributed methods are more popular than centralized methods due to their advantages in reliability, safety, efficiency, and energy consumption. Distributed localization involves designing estimation algorithms to ensure that each agent can estimate its own global position, as precise localization is a prerequisite for multi-agent systems to perform complex tasks. Distributed formation refers to designing control algorithms to drive all agents, such as drones, unmanned vehicles, and unmanned boats, to form a desired formation. However, in the integrated exploration of localization estimation and formation control in multi-agent systems, there are a series of challenges in studying local information exchange and group relationships among agents. Therefore, the integration of localization and formation has become a focal point for current researchers. The "Notice of the State Council on Printing and Distributing the Development Plan for a New Generation of Artificial Intelligence" emphasizes that prioritizing the development of basic theories such as swarm intelligence and autonomous collaborative control is key to overcoming the bottlenecks faced by the new generation of artificial intelligence. Therefore, utilizing the advantages of autonomous decision-making and control in multi-agent systems to achieve integrated autonomous localization and formation control has become an inevitable trend for future development.

## 1.2 Basic Issues of Localization and Formation

Distributed localization and formation control are closely related topics. This section will focus on describing the differences and inherent relationships between localization estimation and formation control.

Consider a multi-agent system in two-dimensional plane, using wheeled robots as an example. Each robot can measure its neighbors using sensors equipped on itself. The variables that can be measured are usually relative distance, bearing, and angle. For localization estimation, the goal is to design a distributed localization estimation algorithm by utilizing locally measured information and communication with neighboring agents, ultimately achieving accurate global position estimation of the robot itself. For formation control, the goal is to design a distributed formation controller using locally measured information, communication with neighboring agents, and a given desired formation to drive the robots to their desired positions.

In short, distributed localization and formation can be viewed as dual problems. Localization involves converging any given virtual coordinates to the robot's true coordinates, while formation involves driving the robot from its current true coordinates to the desired target. The mechanism of both will be detailed using six robots as an example. As shown in Fig. 1.3a, distributed localization designs an estimation algorithm to gradually converge any given estimated position to the real position. Correspondingly, distributed formation control designs a controller to drive the wheeled robots from their current real positions to the desired positions, as shown in Fig. 1.3b. Due to physical hardware limitations, sudden environmental changes, and other uncertainties, there are several critical issues in the research of localization estimation and formation control for multi-agent systems. These issues include transmission delays and losses, random arbitrary distributions, measurement and communication noise, random variations in trajectory length, sensor multiplicative failures, and difficulties in obtaining global information. Achieving precise localization estimation and formation control for multi-agent systems in complex dynamic scenarios is currently a pressing challenge.

### *1.2.1 Transmission Delays and Losses*

Networked multi-agent systems use networks to transmit system signals, which saves wiring costs compared to traditional control systems and has been widely applied in various fields. However, due to hardware limitations, system signals transmitted

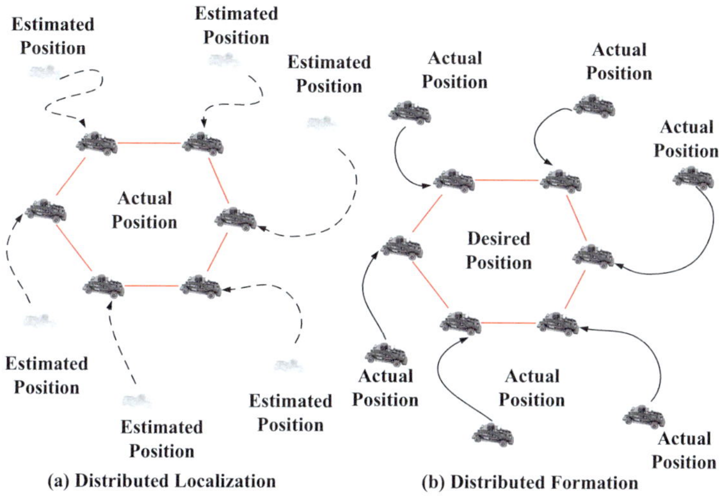


**Fig. 1.3** Examples of distributed localization and distributed formation

over networks inevitably encounter delays and packet losses, which in turn affect system control performance.

Transmission delays and packet losses are important issues that need to be addressed in networked multi-agent systems. The uncertainty of communication delays significantly impacts synchronization events between nodes and includes four main components: sending delay, access delay, propagation delay, and receiving delay. Additionally, due to the unreliability of network transmission, network channels inevitably experience congestion or disconnection, causing data packets to be lost during transmission between agents, preventing them from reaching the target point. Packet loss can generally be categorized into active packet loss and passive packet loss. Networked multi-agent systems typically have high real-time data requirements and can only tolerate a certain degree of packet loss. Once the packet loss rate exceeds a certain threshold, the system becomes unstable.

### *1.2.2 Random Arbitrary Deployment*

Accurate localization is a basic requirement for multi-agent systems to perform different tasks. When an agent cannot directly obtain its own position, it is essential to design a localization estimation method. For agents whose perception variables are relative distance information, a necessary and sufficient condition for the multi-agent system to be localizable is that the communication network topology satisfies global rigidity. Specifically, in a two-dimensional plane, there must be at least three non-collinear leader agents, while in three-dimensional space, there must be at least four non-coplanar leader agents.

Using perceived relative distance information to calculate centroid coordinates to represent the relative positional relationship between agents has a significant advantage in that it expresses agent positions as a pseudo-linear system, with all non-linearity hidden within the measured relative distances. However, this method requires all agents that need to be localized to be within the convex hull formed by the leader agents, which limits its application in real systems. When agents are randomly distributed in a two-dimensional plane or three-dimensional space, the sum of the centroid coordinates calculated from relative distance information exceeds 1, making traditional convex hull algorithms no longer applicable. Developing a distributed localization estimation algorithm to address the localization estimation problem of multi-agent systems under random arbitrary deployment has significant theoretical and practical value.

### 1.2.3 Measurement and Communication Noise

In real-world applications, the perception and transmission of information are inevitably affected by various noise pollutants, usually due to random events occurring during measurement or network transmission. In the study of localization and formation control, measurement noise and communication noise are two important factors that cannot be ignored, as they directly impact system accuracy and stability. Measurement noise primarily originates from errors when sensors acquire data. These errors can arise from various factors, such as low sensor precision, environmental interference, and data processing errors. Communication noise mainly affects information exchange in multi-robot systems or UAV clusters. In formation control, robots or UAVs need real-time communication to coordinate actions and maintain formation. Communication noise can be caused by factors such as signal attenuation, electromagnetic interference, and bandwidth limitations.

Most existing localization and formation control algorithms assume an ideal application environment. Hence, considering random noise in the localization estimation and formation control process, by choosing sensors wisely, optimizing data processing algorithms, enhancing communication capabilities, and introducing redundancy mechanisms, we can reduce the impact of noise on system performance. Improving system accuracy and stability to achieve precise localization and formation control in multi-agent systems is of great importance.

### 1.2.4 Random Variations in Trajectory Length

Random variations in trajectory length refer to the random fluctuations or changes in length or distance traveled when describing an agent's movement or path. This situation can occur in many real-world physical processes, especially those involving randomness or uncertainty. In real-world applications, considering safety issues or the operational capability of controlled systems, external environmental disturbances or hardware faults can cause agents to stop operating prematurely, leading to incomplete trajectory data from the current operation. In other words, the trajectory length of agents is randomly variable, such as in robotic arms and bipedal robot step systems.

During localization estimation, it is expected that agents can maintain a complete trajectory during operation. Random variations in trajectory length cause estimators to be unable to complete normal localization estimation, thus affecting the accuracy and convergence speed of localization. Ensuring accurate localization estimation for dynamic multi-agent systems with random trajectory variations is a topic of significant research value.

### 1.2.5 Sensor Multiplicative Failures

In the field of control, sensors play a crucial role as the main devices for detecting and acquiring information. Reliable sensor operation is a necessary condition for the normal functioning of multi-agent systems and directly impacts system performance. The performance of intelligent sensors that provide measurement and control information directly affects system quality.

Sensor faults are one of the main causes of performance instability and occur more frequently than actuator faults, often due to the complexity of the application environment. Localization estimation and formation control performance depend on the quality of sensor measurement information. Sensor multiplicative faults refer to a type of sensor fault where the actual output value of a sensor is a multiple (non-one) of the normal output signal. Addressing localization estimation and formation control problems under sensor failure conditions is typically challenging because, aside from the perceived sensor information, there is no other feedback information. Therefore, researching compensatory algorithms for perceived variables is crucial for accurate localization estimation and formation control.

### 1.2.6 Difficulty in Obtaining Global Information

In recent years, due to the rapid development of sensors such as optical cameras and LiDAR, bearing information has become easier to obtain and more accurate. The advantage of bearing-based localization and formation control methods is that they can be expressed as a linear dynamic system, making it easier to analyze their global stability.

A technical difficulty in bearing-based localization estimation and formation control for multi-agent systems is that each agent has its local reference frame, which is not aligned with the global reference frame, and the true direction is unknown. It is challenging to directly obtain global bearing information within local reference frames. Therefore, solving the alignment problem of multi-agent systems' orientation under unknown global reference frames is very important.

## 1.3 Overview of Research on Distributed Localization

Over the past thirty years, researchers in both academia and industry have systematically studied distributed localization based on measurement information, proposing many classical localization methods. This section provides a partial overview of existing literature from three perspectives: localization methods based on relative distance, localization methods based on bearing information, and localization methods based on angle information.

### *1.3.1 Localization Methods Based on Relative Distance*

Compared to bearing information and angle information, obtaining relative distance is relatively easier [27]. The measurement techniques for relative distance data can be roughly divided into the following categories: Angle of Arrival (AOA) methods [28–32], Time of Arrival (TOA) methods [33–35], Time Difference of Arrival (TDOA) methods [36–39], and Received Signal Strength (RSS) methods [40–44], among others.

When the perceived variable is relative distance information, a necessary and sufficient condition for the multi-agent system and sensor network to be localizable is that the communication network topology satisfies global rigidity [45–47]. Specifically, in a two-dimensional plane, at least three non-collinear leader agents or anchor nodes are required, while in three-dimensional space, at least four non-coplanar leader agents or anchor nodes are necessary. Anderson et al. [48] were the first to propose the concept of a sequential localization algorithm, which essentially starts from anchor nodes with known global coordinates and localizes the sensor nodes sequentially. Sequential localization is simple and easy to implement, but it requires the sensor network to perform trilateration, which presents certain limitations [49, 50]. Most existing distributed localization methods are based on optimization principles, such as Multidimensional Scaling (MDS) [51–54] and Semidefinite Programming (SDP) [55–58].

It should be noted that the above algorithms exhibit non-convexity, making it impossible to guarantee global convergence of the localization estimation algorithm. Additionally, the selection of initial estimated coordinates can also affect the final localization result. Khan et al. [59] proposed a distributed localization estimation method based on barycentric coordinate representation, addressing the problem of sensor localization estimation in both two-dimensional planes and three-dimensional spaces. A significant advantage of this algorithm is that it represents the sensor's position as a pseudo-linear system, with all non-linearities hidden in the measured relative distances. However, the algorithm requires all sensor nodes to be located within the convex hull formed by the anchor nodes, which limits its application in practical systems. Lin et al. [60] extended Khan's algorithm and proposed a distributed localization estimation method that addresses the localization estimation problem for random deployment sensor networks in two-dimensional planes, removing the convex hull assumption. Furthermore, literature [61] proposed a distributed localization algorithm based on MDS, solving the problem of precise localization of random deployment sensor networks in three-dimensional space. Considering the complex application environment, Khan et al. [62, 63] studied the problem of precise localization of sensor networks with measurement noise. Tian et al. [64] further considered the localization estimation problem for sensor networks with random data packet loss and communication delays.

### *1.3.2 Localization Methods Based on Bearing Information*

In recent years, due to the rapid development of sensors such as optical cameras and LiDAR, bearing information has become easier to obtain and is more accurate. As a result, distributed localization algorithms based on bearing information have gained increasing attention [65–67]. The advantage of bearing-based localization algorithms is that they can be represented as a linear dynamical system, making it easier to analyze their global stability [68]. The localizability of sensor networks or multi-agent systems based on bearing information relies on bearing rigidity theory [69–72], also referred to as parallel rigidity theory. In the case where the global reference frame is known, the localization problem for sensor networks based on bearing has been widely studied [73–75]. Xie et al. [74] solved the distributed localization problem for multi-agent systems in dynamic networks where positions change over time. Lin et al. [75] studied the problem of precise localization in sensor networks under switching topology conditions. In recent years, the problem of distributed localization for sensor networks under local reference frames has also been systematically studied. Literature [76] developed a distributed consensus algorithm that uses relative bearing information to achieve orientation alignment, with the requirement that all initial orientation angles be less than. Ahn et al. [77] removed the initial bearing angle constraint but required the proposed algorithm to perform Gram-Schmidt orthogonalization at each sampling instant. Zhang et al. [78] studied the localization problem of sensor networks based on local orientation information and proposed a distributed orientation estimation and position estimation scheme. Meanwhile, the distributed localization problem for sensor networks under noisy conditions has also been studied [79–82].

### 1.3.3 *Localization Methods Based on Angle Information*

In recent years, distributed localization algorithms based on angle information have begun to attract the attention of researchers [83, 84]. Since the angle between two edges can be calculated by subtracting the bearing values of the two edges measured in the local coordinate system of the sensor, angle-based localization algorithms do not involve the issue of alignment between the local reference frame and the global reference frame. Chen et al. [85] proposed the triangular angle rigidity theory, providing the necessary and sufficient conditions for the global convergence of angle-based localization algorithms in two-dimensional planes, and further extended the results to three-dimensional space [86]. Jing et al. [87, 88], based on angle rigidity theory, introduced the concept of angle fixability and applied the results of angle fixability to solve the localization problem of sensor networks by establishing the connection between angle fixability and angle localizability.

## 1.4 Overview of Research on Distributed Formation Control

Formation control is one of the most active research topics in the field of multi-agent systems, typically aimed at driving multiple agents to achieve specified constraints on their states. Broadly speaking, formation control includes flocking control, where agents move at a uniform velocity while maintaining relative distances, and containment control, where follower agents move within the convex hull formed by leader agents. This section provides an overview of the current state of research on distributed formation control based on agents' sensing capabilities and interaction topology.

Ahn et al. [89] classify distributed formation control into the following categories: formation based on global absolute positions, formation based on relative displacements, formation based on relative distances, and formation based on angles and orientations.

### 1.4.1 *Formation Methods Based on Absolute Positions*

Agents can directly obtain their positions relative to a global reference frame using GPS, Beidou navigation, and other positioning devices. They can actively control their positions to achieve the desired formation described by expected positions relative to the global reference frame. Ren et al. [90, 91] studied formation control for multi-agent systems described by second-order dynamic models and proposed a consensus-based formation control method. This method unifies behavior-based, virtual structure-based, and leader-follower approaches, extending formation control from undirected graphs to directed graphs. Meanwhile, researchers have also conducted systematic in-depth studies on more complex dynamic models, such as higher-order, nonlinear dynamics, and time-varying targets [92–96].

### *1.4.2 Formation Methods Based on Relative Displacement*

Agents actively control their displacement relative to neighboring agents to achieve the desired formation, which is specified by the desired displacement relative to the global reference frame. Literature [97] introduced the consensus concept into formation control problems through coordinate transformations. Most current formation control algorithms are based on consensus algorithms, which have yielded many significant results. Existing displacement-based formation control methods are mainly divided into four categories based on the models used: first-order dynamic models [98–100], second-order dynamic models [101–103], general linear models [104, 105], and non-holonomic constraint dynamic models [106–108].

### *1.4.3 Formation Methods Based on Relative Distance*

Agents actively control the relative distance between themselves and neighboring agents to achieve the desired formation, with the desired formation specified by the expected distances between agents. Distance-based formation control methods do not involve the alignment of local and global reference frames. The challenge lies in designing rigidity control strategies that can ensure global convergence. By utilizing rigidity theory from graph theory [109], researchers have studied distance-based formation controllers and global convergence [110–113]. Ahn et al. [114, 115] proposed a distance-based formation control strategy for agents modeled by single and double integrators in a two-dimensional plane. The proposed control law achieves local asymptotic stability of infinitesimal rigid structures and global asymptotic stability of triangular infinitesimal rigid structures, with all squared distance errors converging exponentially to zero. Anderson et al. [116] explored the problem of tetrahedral formation control for multi-agent systems in three-dimensional space and demonstrated that the designed gradient formation control algorithm based on rigidity theory is almost globally asymptotically stable by studying the system's linearized dynamics.

### 1.4.4 Formation Methods Based on Bearing and Angle

Agents can measure the bearing or angle of neighboring agents in their local reference frame and actively control the relative bearing or angle with their neighbors to achieve the desired orientation or angle. Unlike relative position and distance, the target formation for pure bearing control can be defined using relative bearings obtained through vision sensors or wireless sensor arrays. Over the past decade, formation control algorithms based solely on bearing have attracted significant attention from researchers [117–125]. Due to the relative ease of obtaining bearing information, bearing-based formation control algorithms provide a potential solution for formation control using only onboard sensors.

In two-dimensional planes, some early results on bearing-constrained formation control can be found in the literature [117, 118]. Based on parallel rigidity theory, Bishop et al. [117] introduced the bearing-constrained rigidity matrix and proposed a control algorithm with local asymptotic stability. Although the target formation is defined by relative bearings, relative position measurements are still required. Giordano et al. [118] eliminated the need for relative position measurements by introducing a distributed position estimator. To achieve pure bearing formation control in high-dimensional spaces, Zhao et al. [119] extended bearing rigidity theory to arbitrary dimensions and proposed a control algorithm for infinitesimal bearing rigid formations with almost global asymptotic stability. To further characterize the algebraic properties of bearing rigidity structures, Zhao et al. [120] proposed the bearing Laplacian matrix, which can be used to verify the uniqueness of target formations in any dimension. Based on this powerful tool, Ding et al. [121] designed a new pure bearing control algorithm that ensures global exponential convergence. Since most formation control tasks have time constraints, convergence time is considered an important performance metric. In recent years, researchers have proposed several finite-time and prescribed-time formation control algorithms based on bearing measurements in a global reference frame [122–125].

Eren et al. [126] discussed the feasibility of angle-based formation control methods and provided some preliminary results. Jing et al. [127] proposed an angle rigidity theory to solve formation control problems in two-dimensional planes, designing a distributed formation controller for shape stability and proving the local exponential stability of triangular frame formations. Chen et al. [128–130] conducted systematic research on angle-based formation control problems. Cao et al. [128] proposed an infinitesimal angle rigidity theory for two-dimensional planes and designed a formation stabilization algorithm, achieving angle-rigid formations based solely on angle measurements. Chen et al. [129] discussed the global stability of triangular rigid formations and, inspired by the angle-induced linear equations in triangles, proposed a formation control algorithm that achieves almost global stability of triangular angle-rigid formations using local relative position measurements. Further, Sun et al. [130] ensured the global stability of triangular angle-rigid formations with specified direction and scale by additionally controlling the relative position between a pair of neighboring agents.

## 1.5 Overview of System Stability

### *1.5.1 Graph Theory*

In the problem of distributed localization estimation and formation control for multi-agent systems, the communication topology and relative position relationships between agents can be represented by a directed graph $\mathcal{G}$ [131]. A directed graph $\mathcal{G}$ consists of a finite set of vertex $v$ and a finite set of edge $\varepsilon$, where the vertex set $v = \{v_1, v_2, \ldots, v_n\}$ and the edge set $\varepsilon \subseteq \left\{(v_\alpha, v_\beta) : v_\alpha, v_\beta \in v\right\}$. If $(v_\alpha, v_\beta) \in \varepsilon$, it means agent $v_\alpha$ is a neighbor of agent $v_\beta$, that is, $\mathcal{N}_{v_\alpha} = \left\{v_\beta : (v_\alpha, v_\beta) \in \varepsilon\right\}$. Specifically, agent $v_\beta$ can observe agent $v_\alpha$ and receive information transmitted from agent $v_\alpha$.

An undirected graph can be regarded as a special case of a directed graph. Specifically, for agents $v_\alpha$ and $v_\beta$, if $(v_\alpha, v_\beta) \in \varepsilon$ and $(v_\beta, v_\alpha) \in \varepsilon$, then there is bidirectional communication between agents $v_\alpha$ and $v_\beta$, and they can observe each other. Figure 1.4 illustrate examples of directed and undirected graphs.

Based on directed topologies, the definition of a Henneberg configuration [132] is given.

A directed topology $\mathcal{G} = (v, \varepsilon)$ consists of agents $v_i$, $v_k$ and a directed edge $(v_k, v_i)$. A new directed topology $\mathcal{G}' = \left\{v', \varepsilon'\right\}$ can be generated from graph $\mathcal{G}$ by adding a new agent and splitting a directed edge. Figure 1.5 presents an example of a Henneberg configuration, where the red dots represent leader agents, and the other colored dots represent follower agents.

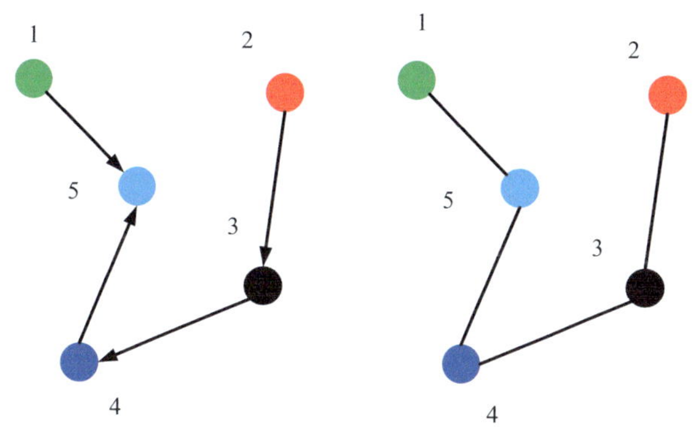


**Fig. 1.4** Examples of directed and undirected graphs

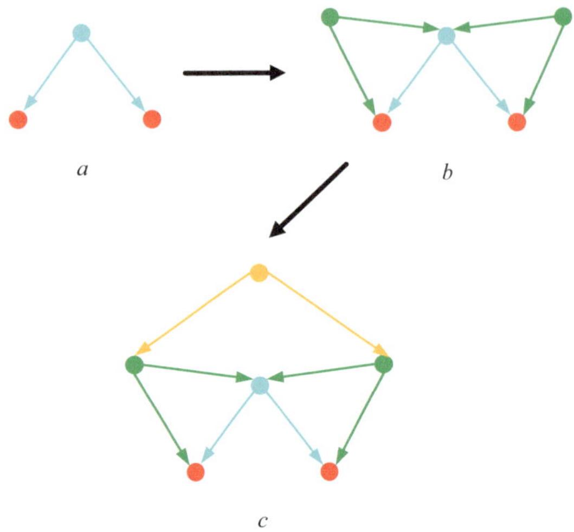


**Fig. 1.5** Example of Henneberg configuration

### 1.5.2 Barycentric Coordinates

When agents use sensors to measure perceived variables as relative distances, barycentric coordinates can be introduced to represent the relative position relationships between agents. The following sections will introduce the methods for calculating barycentric coordinates in both two-dimensional planes and three-dimensional spaces.

As shown in Fig. 1.6, barycentric coordinates are used to describe the position of agent relative to its neighboring agents. In a two-dimensional plane, at least three non-collinear neighboring agents are required to enable agent to estimate its own position. In Fig. 1.6a, agent $l$ is located inside the convex hull formed by neighboring agents $h$, $j$ and $k$. In Fig. 1.6b, agent $l$ is arbitrarily deployment.

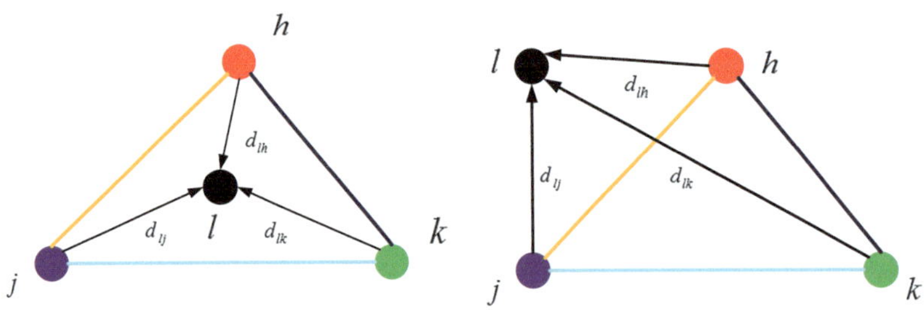


(a) $l$ is located in convex hull (b) $l$ is deployed arbitrarily

**Fig. 1.6** The deployment of agent $l$ relative to agents $h$, $j$ and $k$

The Euclidean coordinates of agent $l$ are defined as $\mathbf{c}_l(t) = [c_{l,1}(t), c_{l,2}(t)]^{\mathrm{T}}$, and the Euclidean coordinates of the other three neighboring agents $h$, $j$ and $k$ are defined as $\mathbf{c}_h(t)$, $\mathbf{c}_j(t)$ and $\mathbf{c}_k(t)$, respectively. The barycentric coordinates of agent $l$ relative to its neighbors $h$, $j$ and $k$ are denoted as $\{a_{l,h}(t), a_{l,j}(t), a_{l,k}(t)\}$. In Fig. 1.6a, the barycentric coordinates satisfy: $a_{l,h}(t) + a_{l,j}(t) + a_{l,k}(t) = 1$, and $\mathbf{c}_l(t) = a_{l,h}(t)\mathbf{c}_h(t) + a_{l,j}(t)\mathbf{c}_j(t) + a_{l,k}(t)\mathbf{c}_k(t)$, where $t$ represents the sampling interval. The relationship between the barycentric coordinates in Fig. 1.6b will be provided in Chap. 2.

The barycentric coordinates can be calculated using the following formula:

$$a_{l,h}(t) = \tfrac{s_{\Delta ljk}(t)}{s_{\Delta hjk}(t)},\ a_{l,j}(t) = \tfrac{s_{\Delta lkh}(t)}{s_{\Delta hjk}(t)},\ a_{l,k}(t) = \tfrac{s_{\Delta lhj}(t)}{s_{\Delta hjk}(t)},$$

where $s$ denotes the area of the corresponding triangle [133].

The area of a triangle can be calculated by the Cayley-Menger determinant [134]. It is given as follows:

$$s^2_{\Delta ljk}(t) = -\tfrac{1}{16}\begin{vmatrix} 0 & 1 & 1 & 1 \\ 1 & 0 & d^2_{lj}(t) & d^2_{lk}(t) \\ 1 & d^2_{jl}(t) & 0 & d^2_{jk}(t) \\ 1 & d^2_{kl}(t) & d^2_{kj}(t) & 0 \end{vmatrix}$$

where $d_{lj}(t)$ represents the relative distance between agents $l$ and $j$ at the $t$th sampling time.

As shown in Fig. 1.7, when only relative distance variables are available, at least four non-coplanar neighboring agents $n$, $h$, $j$ and $k$ are required to enable agent $l$ to estimate its position accurately in a three-dimensional space.

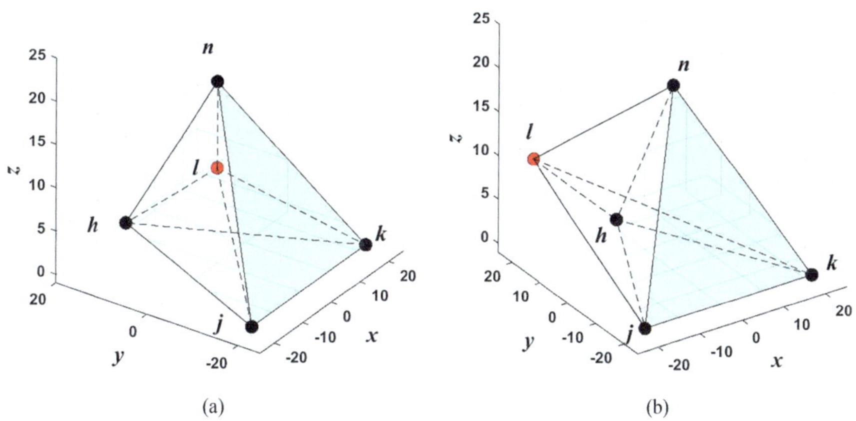


**Fig. 1.7** Example of the deployment of agent $l$ relative to neighbouring agents in 3D space

In Fig. 1.7a, agents $n$, $h$, $j$ and $k$ are leaders with known coordinates, while agent $l$ is a follower whose coordinates need to be estimated. Let $\mathbf{c}_\tau(t)$ be the three-dimensional coordinate vector of agent $\tau$, $\tau \in \{l, n, h, j, k\}$ in the global frame. The barycentric coordinates of agent $l$ relative to its neighbors $n$, $h$, $j$ and $k$ are defined as $\{a_{l,n}(t), a_{l,h}(t), a_{l,j}(t), a_{l,k}(t)\}$. The barycentric coordinates can be calculated using the following formula:

$$a_{l,n}(t) = \frac{V_{lhjk}(t)}{V_{nhjk}(t)}, a_{l,h}(t) = \frac{V_{nljk}(t)}{V_{nhjk}(t)},$$
$$a_{l,j}(t) = \frac{V_{nlkh}(t)}{V_{nhjk}(t)}, a_{l,k}(t) = \frac{V_{nlhj}(t)}{V_{nhjk}(t)},$$

where $V$ represents the volume of the corresponding tetrahedron.

The volume of a tetrahedron can be calculated using the Cayley–Menger determinant:

$$V_{nhjk}^2(t) = \frac{1}{288} \begin{vmatrix} 0 & 1 & 1 & 1 & 1 \\ 1 & 0 & d_{nh}^2(t) & d_{nj}^2(t) & d_{nk}^2(t) \\ 1 & d_{hn}^2(t) & 0 & d_{hj}^2(t) & d_{hk}^2(t) \\ 1 & d_{jn}^2(t) & d_{jh}^2(t) & 0 & d_{jk}^2(t) \\ 1 & d_{kn}^2(t) & d_{kh}^2(t) & d_{kj}^2(t) & 0 \end{vmatrix},$$

where $x, y \in \{n, h, j, k\}$ and $d_{xy}$ represents the relative distance between agents.

As shown in Fig. 1.7a, agent $l$ is located inside the tetrahedron formed by its four neighboring agents $n$, $h$, $j$ and $k$, and the calculated barycentric coordinates satisfy the condition: $a_{l,n}(t) + a_{l,h}(t) + a_{l,j}(t) + a_{l,k}(t) = 1$. The relationship between the barycentric coordinates in Fig. 1.7b will be provided in Chap. 2.

### *1.5.3 Orthogonal Projection Matrix*

As shown in Fig. 1.8, each agent has its own local reference frame, which is not aligned with the global reference frame ${}^g\Sigma$. The perceived variables measured by the agent's sensors are the local bearing angles. Let ${}^i\Sigma$ represent the local reference frame of follower agent $i$, and let $\theta_i \in [0, 2\pi]$ represent the orientation angle of agent $i$ relative to the global frame ${}^g\Sigma$. The displacement vector between agents $i$ and $k$ is defined as $c_{ik} = c_k - c_i$, where $c_k$ and $c_i$ represent the global coordinates of agents $k$ and $i$, respectively. The unit direction vector between agents $i$ and $k$ is $b_{ik} = \frac{c_k - c_i}{\|c_k - c_i\|}$.

The relative bearing vector of agent $k$ as measured by agent $i$ in its local reference frame ${}^i\Sigma$ is denoted as $b_{ik}^i$. Let $R(\theta_i) = \begin{bmatrix} \cos(\theta_i) & -\sin(\theta_i) \\ \sin(\theta_i) & \cos(\theta_i) \end{bmatrix}$ be the rotation matrix of the local reference frame ${}^i\Sigma$ relative to the global reference frame ${}^g\Sigma$. Based on the rotation matrix, the direction vectors of agents $i$ and $k$ in the global frame can be expressed as $b_{ik} = R(\theta_i)\, b_{ik}^i$.

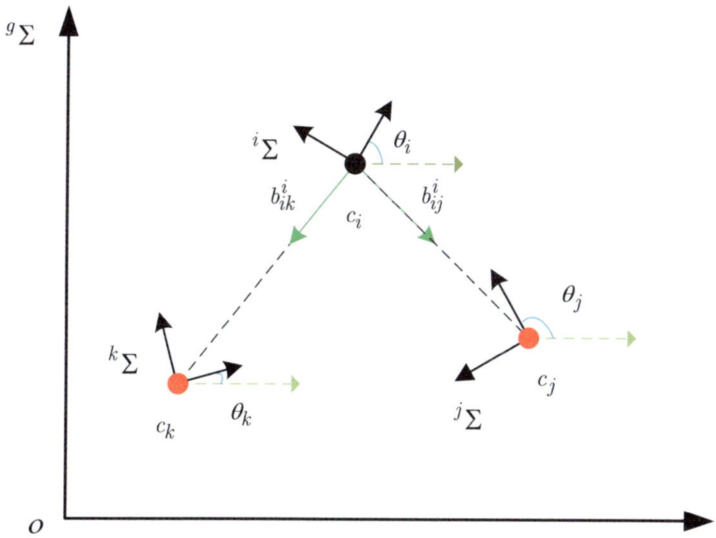


**Fig. 1.8** Example of local bearing measurement

For a unit direction vector $b_{ik}$, the orthogonal projection matrix $Q_{b_{ik}}$ [119] is defined as: $Q_{b_{ik}} = I - b_{ik}b_{ik}^{\mathrm{T}}$, where $I$ is the identity matrix. The orthogonal projection matrix $Q_{b_{ik}}$ can project any vector onto the orthogonal complement of vector $b_{ik}$. It has been proven that the orthogonal projection matrix $Q_{b_{ik}}$ is semi-definite, and satisfies $Q_{b_{ik}}^{\mathrm{T}} = Q_{b_{ik}}$, $Q_{b_{ik}}^2 = Q_{b_{ik}}$ and $Q_{b_{ik}}b_{ik} = 0$.

### *1.5.4 Bearing Localizability*

**Definition 1.1** [135] **Bearing localizability**

A multi-agent network is bearing localizable if and only if the global absolute positions of all follower agents can be uniquely and accurately determined using the global absolute positions of leader agents and the locally measured bearing information.

**Definition 1.2** [136] **Self-localizability of a multi-agent network**

A multi-agent network based on bearing information is called self-localizable if and only if the following conditions are satisfied:

(1) The number of leader agents is greater than or equal to two.
(2) Each follower agent has at least two disjoint paths to the leader agents, ensuring 2-reachability.

**Definition 1.3 Sign function** $sig(\cdot)$:

For a variable $x \in R$, $sig^{\gamma}(x) = sign(x)|x|^{\gamma}$, where $\gamma > 0$, $sign(\cdot)$ is the symbolic function

$$sign\,(x) = \begin{cases} 1, x > 0. \\ 0, x = 0. \\ -1, x < 0. \end{cases}$$

For the scalar, $sig^2(-2) = -4$. For the vector $x \in [x_1, \ldots, x_n]^{\mathrm{T}} \in R^n$, $sig^{\gamma}(x) = \left[sign(x_1)|x_1|^{\gamma}, \ldots, sign(x_n)|x_n|^{\gamma}\right]^{\mathrm{T}}$. For example, if $A = \begin{bmatrix} 1 & -3 \\ 2 & 0 \end{bmatrix}$, then $sig^2(A) = \begin{bmatrix} 1 & -9 \\ 4 & 0 \end{bmatrix}$.

### 1.5.5 Prescribed-Time Stability Theory

**Lemma 1.1** *[137]* ***Prescribed-Time Stability Criterion***

*Consider a continuously differentiable function* $y : R \to R_{\geq 0}$. *If it satisfies the following condition:*

$\dot{y}(t) \leq -a_1 y - b_1 \frac{\dot{\mu}}{\mu} y,$

*where* $t \in [t_0, \infty)$, $a_1, b_1 \in R_{>0}$. *The time-varying gain function*

$\mu(t) = \begin{cases} \frac{T^h}{(t_0+T-t)^h}, t \in [t_0, t_0+T) \\ 1, \quad t \in [t_0+T, \infty) \end{cases}$, $\dot{\mu}(t) = \begin{cases} \frac{h}{T}\mu^{\left(1+\frac{1}{h}\right)}, t \in [t_0, t_0+T) \\ 0, \quad t \in [t_0+T, \infty) \end{cases}$, $h > 0$ *and* $T$ *is the specified convergence time, then the function* $y(t)$ *satisfies:*

$$y(t) \begin{cases} \leq \mu^{-b_1} e^{-a_1(t-t_0)} y(t_0), t \in [t_0, t_0+T) \\ \equiv 0, \quad t \in [t_0+T, \infty) \end{cases}$$

*That is, the function* $y(t)$ *converges to zero within the specified time* $T$.

## References


1. Seiler, P., Pant, A., Hedrick, J.: A systems interpretation for observations of bird V-formations. J. Theor. Biol. **221**(2), 279–287 (2003)
2. Hu, H., Liang, J., Wen, G., et al.: A survey of development on swarming behavior for multi-agent systems. J. Nanjing Univ. Inf. Sci. Technol. (Natural Science Edition), **10**(4), 415–421 (2018)
3. Bonabeau, E., Dorigo, M., Theraulaz, G.: Swarm Intelligence: From Natural to Artificial Systems. Oxford University Press (1999)
4. Seeley, T., Camazine, S., Sneyd, J.: Collective decision-making in honey bees: how colonies choose among nectar sources. Behav. Ecol. Sociobiol. **28**(4), 277–290 (1991)
5. Hewitt, C.: Viewing control structures as patterns of passing messages. Artif. Intell. **8**(3), 323–364 (1977)
6. Sun, M., Ren, L., Liu, J., et al.: Dynamic event-triggered fixed-time consensus control of multi-agent systems under switching topologies. Acta Autom. Sinica **49**(06), 1295–1305 (2023)
7. Wu, M., Zhang, H., Yan, H., et al.: Cooperative output regulation for asynchronously switched multi-agent systems. Acta Autom. Sinica **43**(05), 735–742 (2017)
8. Wang, H., Yu, Y., Jiang, Y.: Review of the progress of communication-based multi-agent reinforcement learning (in Chinese). SCIENTIA SINICA Inf. **52**(05), 742–764 (2022)

9. Gunathillake, A., Huang, H., Savkin, A.: Sensor-network-based navigation of a mobile robot for extremum seeking using a topology map. IEEE Trans. Industr. Inf. **15**(7), 3962–3972 (2019)
10. Li, Y., Xing, S., Zhang, Y.: Maritime search task allocation with multi-agent concept. Navig. China **41**(03), 91–100 (2018)
11. Curtis, S., Rilee, M., Clark, P., et al.: Use of swarm intelligence in spacecraft constellations for the resource exploration of the asteroid belt. In: Proceedings of International Workshop on Satellite Constellations and Formation Flying, pp. 24–26, Pisa, Italy (2003)
12. Chen, Y., Ma, X., Bai, G., et al.: Multi-autonomous underwater vehicle formation control and cluster search using a fusion control strategy at complex underwater environment. Ocean Eng. **216**, 108048 (2020)
13. Rengaraju, P., Pandian, S., Lung, C.: Communication networks and non-technical energy loss control system for smart grid networks. In: Proceedings of IEEE Innovative Smart Grid Technologies-Asia, pp. 418–423, Kuala Lumpur, Malaysia (2014)
14. Liu, B.: Two-layer path planning algorithm of UAV and UGV in uncertain environment. Harbin Institute of Technology (2022)
15. Pu, M., Li, G.: Attitude control and simulation analysis of AGV jacking-up platform. Port Oper. **15**(16), 14–28 (2019)
16. Sun, J., Su, W.: Optimal design of container AGV suspension system based on tire wear. Port Oper. **21**(56), 32–45 (2019)
17. Liu, J., Zhang, H., Wang, Z.: Cooperative output regulation of heterogeneous multi-agent systems by self-triggered. Acta Autom. Sinica **45**(10), 1893–1902 (2019)
18. Ren, W., Beard, R.W., Atkins, E.: Information consensus in multivehicle cooperative control. IEEE Control Syst. Mag. **27**(2), 71–82 (2007)
19. Zhou, F., Wu, Y.: Consensus tracking algorithms with directed network. Acta Autom. Sinica **41**(1), 180–185 (2015)
20. Min, H., Liu, Y., Wang, S., et al.: An overview on coordination control problem of multi-agent system. Acta Autom. Sinica **38**(10), 1557–1570 (2012)
21. Ouammi, A., Zejli, D.: Centralized controller for the optimal operation of a cooperative cluster of connected microgrids powered multi-greenhouses. Sustain. Comput. Inf. Syst. **33**, 100641 (2022)
22. Cui, J., Lu, Q., Zhong, H., et al.: A load-balancing mechanism for distributed SDN control plane using response time. IEEE Trans. Netw. Serv. Manage. **15**(4), 1197–1206 (2018)
23. Moses, R., Krishnamurthy, D., Patterson, R.: A self-localization method for wireless sensor networks. EURASIP J. Adv. Signal Process. **4**, 348–358 (2003)
24. Patwari, N., Hero, A., Perkins, M., et al.: Relative location estimation in wireless sensor networks. IEEE Trans. Signal Process. **51**(8), 2137–2148 (2003)
25. Zheng, W., Xu, Y., Luo, D.: Distributed hierarchical formation-containment control of multiple quadrotor UAV systems. J. Beijing Univ. Aeronaut. Astronaut. **48**(06), 1091–1105 (2022)
26. Diao, Y., Lin, Z., Fu, M.: A barycentric coordinate based distributed localization algorithm for sensor networks. IEEE Trans. Signal Process. **62**(18), 4760–4771 (2014)
27. Mao, G., Fidan, B., Anderson, B.: Wireless sensor network localization techniques. Comput. Netw. **51**, 2529–2553 (2007)
28. Yang, Y.: The research on Doppler frequency shift and angle of arrival measurement system based on microwave photonics. Jilin University (2023)
29. Rappaport, T., Reed, J., Woerner, B.: Position location using wireless communications on highways of the future. IEEE Commun. Mag. **34**(10), 33–41 (1996)
30. Biedka, T., Reed, J., Woerner, B.: Direction finding methods for CDMA systems. In: Proceedings of the Thirtieth Asilomar Conference on Signals, Systems and Computers, pp. 637–641, Pacific Grove, USA (1996)
31. Bliss, D., Forsythe, K.: Angle of arrival estimation in the presence of multiple access interference for CDMA cellular phone systems. In: Proceedings of IEEE Sensor Array and Multichannel Signal Processing Workshop, pp. 408–412, Cambridge, USA (2000)

32. Ziskind, I., Wax, M.: Maximum likelihood localization of multiple sources by alternating projection. IEEE Trans. Acoust. Speech Signal Process. **36**(10), 1553–1560 (1988)
33. Venkatraman, S., Caffery, J., Heung-Ryeol, Y.: A novel TOA location algorithm using LoS range estimation for NLOS environments. IEEE Trans. Veh. Technol. **53**(5), 1515–1524 (2004)
34. Hong, G.: TOA-second-order Programming-based Localization for wireless sensor networks. Fire Control Command Control **45**(09), 63–71 (2020)
35. McCrady, D., Doyle, L., Forstrom, H., et al.: Mobile ranging using low-accuracy clocks. IEEE Trans. Microw. Theory Tech. **48**(6), 951–958 (2000)
36. Carter, G.: Time delay estimation for passive sonar signal processing. IEEE Trans. Acoust. Speech Signal Process. **29**(3), 463–470 (1981)
37. Gardner, W., Chen, C.: Signal-selective time-difference-of-arrival estimation for passive location of man-made signal sources in highly corruptive environments. I. Theory and method. IEEE Trans. Signal Process **40**(5), 1168–1184 (1992)
38. Zhang, H.: Research on directional target localization technology based on hybrid measurement. Xidian Ubiversity (2022)
39. Li, C., Zhuang, W.: Nonline-of-sight error mitigation in mobile location. IEEE Trans. Wirel. Commun. **4**(2), 560–573 (2005)
40. Bahl, P., Padmanabhan, V.: RADAR: an in-building RF-based user location and tracking system. In: Proceedings of IEEE INFOCOM, vol. 2, pp. 775–784, Tel Aviv, Israel (2000)
41. Yu, Z.: Research on indoor navigation and positioning algorithm based on RSSI ranging. Henan Polytechnic University (2023)
42. Krishnan, P., Krishnakumar, A., Ju, W., et al.: A system for LEASE: location estimation assisted by stationary emitters for indoor RF wireless networks. In: Proceedings of IEEE INFOCOM, vol. 2, pp. 1001–1011, Hong Kong, China (2004)
43. Roos, T., Myllymaki, P., Tirri, H.: A statistical modeling approach to location estimation. IEEE Trans. Mob. Comput. **1**(1), 59–69 (2002)
44. Ray, S., Lai, W., Paschalidis, I.: Deployment optimization of sensor net-based stochastic location-detection systems. In: Proceedings of IEEE INFOCOM, Miami, USA **4**, 2279–2289 (2005)
45. Eren, T., Goldenberg, D., Whiteley, W., et al.: Rigidity, computation, and randomization in network localization. In: Proceedings of Annual Joint Conference of the IEEE Computer and Communications Societies, pp. 2673–2684, Hong Kong, China (2004)
46. Hendrickson, B.: Conditions for unique graph realizations. SIAM J. Comput. **21**, 65–84 (1992)
47. Jackson, B., Jordán, T.: Connected rigidity matroids and unique realizations of graphs. J. Comb. Theory, Ser. B **94**, 1–29 (2005)
48. Fang, J., Cao, M., Morse, A., et al.: Sequential localization of sensor networks. SIAM J. Control Optim. **48**(1), 321–350 (2009)
49. Cao, M., Anderson, B., Morse, A.: Sensor network localization with imprecise distances. Syst. Control Lett. **55**(11), 887–893 (2006)
50. Anderson, B., Belhumeur, P., Eren, T., et al.: Graphical properties of easily localizable sensor networks. Wirel. Netw. **15**(2), 177–191 (2009)
51. Costa, J., Patwari, N., Hero, A.: Distributed weighted-multidimensional scaling for node localization in sensor networks. ACM Trans. Sens. Netw. **2**(1), 39–64 (2006)
52. Chan, F., So, H.C.: Efficient weighted multidimensional scaling for wireless sensor network localization. IEEE Trans. Signal Process. **57**(11), 4548–4553 (2009)
53. Cox, T., Cox, M.: Multidimensional Scaling. Chapman & Hall, London (1994)
54. Luo, H., Li, J., Zhu, Z., et al.: Distributed multidimensional scaling with relative error-based neighborhood selection for node localization in sensor networks. In: Proceedings of IEEE International Conference on Integration Technology, pp. 735–739, Shenzhen, China (2007)
55. Biswas, P., Lian, T., Wang, T., et al.: Semidefinite programming-based algorithms for sensor network localization. ACM Trans. Sens. Netw. **2**, 188–220 (2006)
56. Du, Y., Wei, P., Zhang, H.: Semidefinite programming approaches for source localization problems. In: Proceedings of IEEE International Conference on Communication Problem-Solving, pp. 339–342, Beijing, China (2014)

57. Wang, Z., Zhang, H., Lu, T., et al.: Cooperative RSS-based localization in wireless sensor networks using relative error estimation and semidefinite programming. IEEE Trans. Veh. Technol. **68**(1), 483–497 (2019)
58. Lui, K., Ma, W.K., So, H., et al.: Semi-definite programming algorithms for sensor network node localization with uncertainties in anchor positions and/or propagation speed. IEEE Trans. Signal Process. **57**(2), 752–763 (2009)
59. Khan, U., Kar, S., Moura, J.: Distributed sensor localization in random environments using minimal number of anchor nodes. IEEE Trans. Signal Process. **57**(5), 2000–2016 (2009)
60. Diao, Y., Lin, Z., Fu, M.: A barycentric coordinate based distributed localization algorithm for sensor networks. IEEE Trans. Signal Process. **62**(18), 4760–4771 (2014)
61. Han, T., Lin, Z., Zheng, R., et al.: A barycentric coordinate based approach to three-dimensional distributed localization for wireless sensor networks. In: Proceedings of IEEE International Conference on Control & Automation, pp. 600–605, Ohrid, Macedonia (2017)
62. Khan, U., Kar, S., Moura, J.: Diland: an algorithm for distributed sensor localization with noisy distance measurements. IEEE Trans. Signal Process. **58**(3), 1940–1947 (2010)
63. Wu, F., Tian, Y., Wang, B.: Distributed localization in sensor networks with noisy distance measurements. In: Proceedings of Chinese Control Conference, pp. 8917–8922, Dalian, China (2017)
64. Huang, X.: Distributed localization algorithms in uncertain network environment. Southeast University (2018)
65. Gao, B.: Research on target localization and single sensor trajectory planning with constraints based on bearing-only measurements. Central South University (2022)
66. Hu, B., Zhang, H.: Bearing-based localization of multi-agent system with event-triggered strategy. Control Theory Appl. **38**(11), 1845–1854 (2021)
67. Piovan, G., Shames, I., Fidan, B., et al.: On frame and orientation localization for relative sensing networks. Automatica **49**(1), 206–213 (2013)
68. Zhao, S., Zelazo, D.: Bearing rigidity theory and its applications for control and estimation of network systems: Life beyond distance rigidity. IEEE Control Syst. Mag. **39**(2), 66–83 (2019)
69. Zelazo, D., Franchi, A., Giordano, P.: Rigidity theory in SE(2) for unscaled relative position estimation using only bearing measurements. In: Proceedings of European Control Conference, pp. 2703–2708, Strasbourg, France (2014)
70. Zhao, S., Zelazo, D.: Bearing rigidity and almost global bearing only formation stabilization. IEEE Trans. Autom. Control **61**(5), 1255–1268 (2016)
71. Lin, Z., Han, T., Zheng, R., et al.: Distributed localization for 2-D multi-agent systems with bearing-only measurements under switching topologies. IEEE Trans. Signal Process. **64**(3), 6345–6359 (2016)
72. Tron, R., Carlone, L., Dellaert, F., et al.: Rigid components identification and rigidity control in bearing-only localization using the graph cycle basis. In: Proceedings of American Control Conference, pp. 3911–3918, Chicago, USA (2015)
73. Zhu, G., Hu, J.: A distributed continuous-time algorithm for network localization using angle-of-arrival information. Automatica **50**, 53–63 (2014)
74. Fang, X., Xie, L., Li, X.: Distributed localization in dynamic networks via complex Laplacian. Automatica **151**, 110915 (2023)
75. Lin, Z., Han, T., Zheng, R., et al.: Distributed localization for 2-D sensor networks with bearing-only measurements under switching topologies. IEEE Trans. Signal Process. **64**(23), 6345–6359 (2016)
76. Oh, K., Ahn, H.: Formation control and network localization via orientation alignment. IEEE Trans. Autom. Control **59**(2), 540–545 (2014)
77. Trinh, M., Lee, B., Ye, M., et al.: Bearing-based formation control and network localization via global orientation estimation. In: Proceedings of IEEE Conference on Control Technology and Applications, pp. 1084–1089, Copenhagen, Denmark (2018)
78. Wang, Z., Chang, Y., Zhang, H., et al.: Bearing-only distributed localization for multi-agent systems with complex coordinates. Inf. Sci. **626**, 837–850 (2023)

79. Shames, I., Bishop, A., Anderson, B.: Analysis of noisy bearing-only network localization. IEEE Trans. Autom. Control **58**(1), 247–252 (2013)
80. Wang, B., Tian, Y.: Distributed network localization: accurate estimation with noisy measurement and communication information. IEEE Trans. Signal Process. **66**(2), 5927–5940 (2018)
81. Bae, Y., Kwon, S., Lim, Y., et al.: Distributed bearing-based formation control and network localization with exogenous disturbances. Int. J. Robust Nonlinear Control **32**(11), 6556–6573 (2022)
82. Luo, J., Xue, C., Peng, D.: Robust bearing-only localization using total least absolute residuals optimization. Complexity 1–13 (2020)
83. Chen, L., Xie, L., Li, X., et al.: Simultaneous localization and formation using angle-only measurements in 2D. Automatica **146**, 110605 (2022)
84. Hai, N.: Research on bearings-only single station passive location and tracking. University of Electronic Science and Technology of China (2016)
85. Chen, L.: Triangular angle rigidity for distributed localization in 2D. Automatica **143**, 110414 (2022)
86. Chen, L., Cao, K., Xie, L., et al.: 3-D network localization using angle measurements and reduced communication. IEEE Trans. Signal Process. **70**, 2402–2415 (2022)
87. Jing, G., Wan, C., Dai, R.: Angle-based sensor network localization. IEEE Trans. Autom. Control **67**(2), 840–855 (2022)
88. Jing, G., Wan, C., Dai, R.: Angle fixability and angle-based sensor network localization. In: Proceedings of IEEE Conference on Decision and Control, pp. 7899–7904, Nice, France (2019)
89. Oh, K., Park, M., Ahn, H.: A survey of multi-agent formation control. Automatica **53**, 424–440 (2015)
90. Ren, W., Atkins, E.: Distributed multi-vehicle coordinated control via local information exchange. Int. J. Robust Nonlinear Control **17**(10–11), 1002–1033 (2007)
91. Ren, W.: Consensus strategies for cooperative control of vehicle formations. IET Control Theory Appl. **1**(2), 505–512 (2007)
92. Fax, A., Murray, R.M.: Information flow and cooperative control of vehicle formations. IEEE Trans. Autom. Control **49**(9), 1465–1476 (2004)
93. Dong, W., Farrell, J.: Cooperative control of multiple nonholonomic mobile agents. IEEE Trans. Autom. Control **53**(6), 1434–1448 (2008)
94. Zhi, H., Chen, L., Li, C., et al.: Leader-follower affine formation control of second-order nonlinear uncertain multi-agent systems. IEEE Trans. Circuit Syst. II Express Briefs **68**(12), 3547–3551 (2021)
95. Wang, R., Dong, X., Li, Q., et al.: Distributed time-varying output formation control for general linear multiagent systems with directed topology. IEEE Trans. Control Netw. Syst. **6**(2), 609–620 (2019)
96. Chen, L., Mei, J., Li, C., et al.: Distributed leader-follower affine formation maneuver control for high-order multiagent systems. IEEE Trans. Autom. Control **65**(11), 4941–4948 (2020)
97. Saber, R., Murray, R.: Consensus problems in networks of agents with switching topology and time-delays. IEEE Transa. Autom. Control **49**(9), 1520–1533 (2004)
98. Han, Z., Guo, K., Xie, L., et al.: Integrated relative localization and leader-follower formation control. IEEE Trans. Autom. Control **64**(1), 20–34 (2018)
99. Lee, B., Ahn, H.: Distributed formation control via global orientation estimation. Automatica **73**, 125–129 (2016)
100. Chen, F., Dimarogonas, D.: Leader-follower formation control with prescribed performance guarantees. IEEE Trans. Control Netw. Syst. **8**(1), 450–461 (2021)
101. Han, T., Lin, Z., Xu, W., et al.: Three-dimensional formation merging control of second-order agents under directed and switching topologies. In: Proceedings of IEEE International Conference on Control Automation, pp. 225–230, Taichung, Taiwan (2014)
102. Rezaee, H., Abdollahi, F.: Pursuit formation of double-integrator dynamics using consensus control approach. IEEE Trans. Industr. Electron. **62**(7), 4249–4256 (2014)

103. Yang, J., Cao, G., Chen, W., et al.: Finite-time formation control of second-order linear multi-agent systems with relative state constraints: a barrier function sliding mode control approach. IEEE Trans. Circ. Syst. II Express Briefs **69**(3), 1253–1256 (2022)
104. Wen, G., Duan, Z., Ren, W., et al.: Distributed consensus of multi-agent systems with general linear node dynamics and intermittent communications. Int. J. Robust Nonlinear Control **24**(16), 2438–2457 (2014)
105. Zuo, S., Song, Y., Lewis, F., et al.: Time-varying output formation containment of general linear homogeneous and heterogeneous multiagent systems. IEEE Trans. Control Netw. Syst. **6**(2), 537–548 (2019)
106. Dimarogonas, D., Kyriakopoulos, K.: A connection between formation infeasibility and velocity alignment in kinematic multi-agent systems. Automatica **44**(10), 2648–2654 (2008)
107. Aloisi, D., Cristofaro, A.: Consensus and formation control of unicycle-like robots with discontinuous communication protocols. In: Proceedings of European Control Conference, pp. 1055–1060, London, United Kingdom (2022)
108. Tran, Q., Ahn, H.: Distributed formation control of mobile agents via global orientation estimation. IEEE Trans. Control Netw. Syst. **7**(4), 1654–1664 (2020)
109. Asimow, L., Roth, B.: The rigidity of graphs. Trans. Am. Math. Soc. **245**, 279–289 (1978)
110. Anderson, B., Yu, C., Fidan, B., et al.: Rigid graph control architectures for autonomous formations. IEEE Control Syst. Mag. **28**, 48–63 (2008)
111. Tian, Y., Wang, Q.: Global stabilization of rigid formations in the plane. Automatica **49**, 1436–1441 (2013)
112. Zelazo, D., Franchi, A., Giordano, P.: Distributed rigidity maintenance control with range-only measurements for multi-robot systems. Int. J. Robot. Res. **34**(1), 105–128 (2015)
113. Sun, Z., Mou, S., Deghat, M., et al.: Finite time distance-based rigid formation stabilization and flocking. IFAC Proc. Vol. **47**(3), 9183–9189 (2014)
114. Oh, K., Ahn, H.: Formation control of mobile agents based on inter-agent distance dynamics. Automatica **47**(10), 2306–2312 (2011)
115. Oh, K., Ahn, H.: Distance-based undirected formations of single-integrator and double-integrator modeled agents in n-dimensional space. Int. J. Robust Nonlinear Control **24**, 1809–1820 (2014)
116. Park, M., Sun, Z., Anderson, B., et al.: Stability analysis on four agent tetrahedral formations. In: Proceedings of IEEE Conference on Decision and Control, pp. 631–636, Los Angeles, USA (2014)
117. Bishop, A., Shames, I., Anderson, B.: Stabilization of rigid formations with direction-only constraints. In: Proceedings of IEEE Conference on Decision and Control and European Control Conference, pp. 746–752, Orlando, USA (2011)
118. Franchi, A., Giordano, P.: Decentralized control of parallel rigid formations with direction constraints and bearing measurements. In: Proceedings of IEEE Conference on Decision and Control, pp. 5310–5317, Maui, USA (2012)
119. Zhao, S., Zelazo, D.: Bearing rigidity and almost global bearing-only formation stabilization. IEEE Trans. Autom. Control **61**(5), 1255–1268 (2016)
120. Zhao, S., Zelazo, D.: Localizability and distributed protocols for bearing-based network localization in arbitrary dimensions. Automatica **69**, 334–341 (2016)
121. Zhao, S., Li, Z., Ding, Z.: Bearing-only formation tracking control of multiagent systems. IEEE Trans. Autom. Control **64**(11), 4541–4554 (2019)
122. Zhao, S., Lin, F., Peng, K., et al.: Finite-time stabilization of cyclic formations using bearing-only measurements. Int. J. Control **87**(4), 715–727 (2014)
123. Trinh, M., Mukherjee, D., Zelazo, D., et al.: Finite-time bearing only formation control. In: Proceedings of IEEE Conference on Decision and Control, pp. 1578–1583, Melbourne, Australia (2017
124. Tran, Q., Trinh, M.H., Zelazo, D., et al.: Finite time bearing-only formation control via distributed global orientation estimation. IEEE Trans. Control Netw. Syst. **6**(2), 702–712 (2019)

125. Li, Z., Tnunay, H., Zhao, S., et al.: Bearing-only formation control with prespecified convergence time. IEEE Trans. Cybern. **52**(1), 620–629 (2022)
126. Eren, T., Whiteley, W., Morse, A., et al.: Sensor and network topologies of formations with direction, bearing, and angle information between agents. In: Proceedings of IEEE Conference on Decision and Control, pp. 3064–3069, Maui, USA (2003)
127. Jing, G., Zhang, G., Lee, H., et al.: Angle-based shape determination theory of planar graphs with application to formation stabilization. Automatica **105**, 117–129 (2019)
128. Chen, L., Cao, M., Li, C.: Angle rigidity and its usage to stabilize multi-agent formations in 2D. IEEE Trans. Autom. Control **66**(8), 3667–3681 (2021)
129. Chen, L., Yang, Q., Shi, M., et al.: Stabilizing angle rigid formations with prescribed orientation and scale. IEEE Trans. Ind. Electron. **69**(11), 11654–11664 (2022)
130. Chen, L., Sun, Z.: Globally stabilizing triangularly angle rigid formations. IEEE Trans. Autom. Control **68**(2), 1169–1175 (2023)
131. Godsil, C., Royle, G.: Algebraic Graph Theory. Springer (2001)
132. Trinh, M., Zhao, S., Sun, Z., et al.: Bearing based formation control of a group of agents with leader-follower structure. IEEE Trans. Autom. Control **64**(2), 598–613 (2019)
133. Coxeter, H.: Introduction to Geometry. John Wiley and Sons, Inc. (1969)
134. Sippl, M., Scheraga, H.: Cayley-Menger coordinates. Natl. Acad. Sci. **83**(8), 2283–2287 (1986)
135. Barooah, P., Hespanha, J.: Estimation on graphs from relative measurements. IEEE Control Syst. Mag. **27**(4), 57–74 (2007)
136. Lin, Z., Fu, M., Diao, Y.: Distributed self-localization for relative position sensing networks in 2D space. IEEE Trans. Signal Process. **63**(14), 3751–3761 (2015)
137. Li, Z., Tnunay, H., Zhao, S., et al.: Bearing only formation control with prespecified convergence time. IEEE Trans. Cybern. **52**, 620–629 (2022)

# Chapter 2
# Distributed Localisation for MASs Under Barycentric Coordinate Representation

This chapter investigates the real-time localization problem of dynamic multi-agent systems with repetitive operation characteristics under directed graph. A distributed localization estimation algorithm based on iterative learning is proposed. The barycentric coordinates calculated based on the relative distance are used to estimate the real coordinates of the agent. Different from the traditional estimation methods along the time axis, the proposed method utilizes the information of iteration axis simultaneously. In this method, the current estimation coordinates is updated by using the estimation coordinates of the same sampling time in previous iteration, the estimation accuracy is improved and the velocity constraint is removed. Additionally, the real-time localization problem of dynamic multi-agent systems under arbitrary deployment is concerned. An improved distributed localization estimation algorithm with signed coefficients based on iterative learning is proposed. Meanwhile, the results are also extended to the localization estimation of multi-agent systems with arbitrary deployment in three-dimensional space. By introducing Richardson iteration and infinite norm, the global asymptotic convergence of the proposed methods is guaranteed. Finally, numerical simulations and the Qbot-2e robot experiment are provided to show the effectiveness and validity of the obtained results.

## 2.1 Introduction

Precise localization is the basic requirement of multi-agent systems for different tasks. It is crucial to design a localization estimation algorithm when an agent cannot obtain its own position directly. In recent years, localization estimation has received more and more attention.

According to the different detection variables used in estimation algorithm, the existing localization methods can be divided into three categories: distance-based

 
Y. Lv et al., *Distributed Localization and Formation Control of Multi-Agent Systems with Local Sensing Information*, https://doi.org/10.1007/978-981-96-6015-5_2

estimation methods , bearing-based estimation methods, and angel-based estimation methods. In recent years, sensor localization estimation algorithm has been extensively and systematically researched. However, it can be found that most of the literature are focused on the accurate localization in static networks. For example, Khan et al. [1] propose a distributed iterative localization estimation algorithm based on barycentric coordinates, which guarantees the global convergence. In [2], an extended distributed computation scheme of coordinates is proposed to relaxes the assumption that the sensor node in a convex hull formed by its neighboring nodes. In [3], a distributed localization scheme based on bearing-measurement is proposed to solve the sensor node localization with switching topology, and the common orientation of local coordinates is avoided. Chen et al. [4] proposed a triangular angular rigidity theory to derive a sufficiently necessary condition for global convergence of angle-based localization algorithms in the 2D plane.

Noted that dynamic is the inherent property of the multi-agent systems in practical applications. Real-time and accurate localization in dynamic networks is an important premise for multi-agent systems to perform different tasks. In recent years, some pioneering research results have come out. In [5, 6], the estimation method based on bearing measurement is proposed to solve the localization problem of slow-moving agents. It is worth noting that the above mentioned works estimate the position coordinates along the time axis. Slow estimation speed is the common requirement for this kind of method, which leads to the problem that it is only suitable for the scene with slow changing properties. The real-time localization of dynamic multi-agent systems with fast changing characteristics is an important and challenging work. Therefore, it is critical to design a real-time localization estimation algorithm for dynamic multi-agent systems.

Note that some dynamic multi-agent systems show the characteristics of repeated operation, such as, power line inspection UAV [7], transportation robot [8], earth satellite [9] and so on, which perform task repeatedly. Thus, the multi-agent systems has two-dimensional dynamics: time dimensional dynamics and repetition dimensional dynamics. Arimoto et al. propose the iterative learning control in 1984 [10]. The operating principle of the iterative learning control is to learn from previous trials, which have shown excellent ability in solving transient real-time response and tracking problems. Motivated by the idea of iterative learning control, we apply two-dimensional learning to the localization estimation of dynamic multi-agent systems. In this framework, the estimated coordinates value of the last operation is used to update the current one to achieve real-time localization. Thus, the estimation speed is increased and the real-time localization is guarant eed.

In a word, a distributed localization estimation algorithm based on iterative learning is proposed in this chapter, which can solve the real-time accurate localization estimation problem of dynamic multi-agent systems with repetitive operation characteristics under directed graph. The main contributions can be summarized as follows:

(1) Different from the existing literature, the repeated dynamic properties of the multi-agent systems have been used in this study.

(2) Different from the works [5, 6], the real-time localization problem of the multi-agent systems without velocity constraint under directed graph is considered.
(3) A distributed localization method based on iterative learning is proposed. The real-time localization of the multi-agent systems with arbitrary deployment is realized.

The rest of this chapter is arranged as follows: Sect. 2.2 shows the problem description. The design and convergence analysis of the estimator in 2D space are introduced in Sect. 2.3. Section 2.4 extends the results to the multi-agent systems with arbitrary deployment in three-dimensional space. In Sect. 2.5, numerical simulation and Qbot-2e robot experiment are provided. Finally, Sect. 2.6 is the conclusion.

## 2.2 Problem Formulation

Consider a dynamic system with $N$ agents with directed graph, in which $m$ agents are leaders, marked $1, \ldots, m$, and the other agents are followers, marked $m+1, \ldots, N$. In the global reference frame, the real position of the agent $l$ is represented by $\mathbf{c}_l(t) = [c_{l,1}(t), c_{l,2}(t)]^{\mathrm{T}}$, where $c_{l,1}(t)$ and $c_{l,2}(t)$ denote the horizontal and vertical coordinates, respectively. The leader agent can obtain its own coordinate value directly, while the follower agents can only obtain their coordinate value indirectly through communication with the leaders. The communication between agents is unidirectional and its communication topology can be represented as $\mathcal{G} = (v, \varepsilon)$. Each agent is able to measure the relative distance data to its neighbouring agents.

The distributed localization estimation problem for multi-agent systems in 2D plane and 3D space is considered respectively. The goal of this chapter is to develop an effective distributed localization estimation scheme for two different application scenarios, namely, convex hull and arbitrary deployment. The proposed algorithm can ensure the estimation error $\boldsymbol{e}_l(t)$ equal to zero, where $\mathbf{e}_l(t) = \hat{\mathbf{c}}_l(t) - \mathbf{c}_l^*(t)$, $l \in \{m+1, \ldots, N\}$, $\hat{\mathbf{c}}_l(t)$ denotes the estimation value and $\mathbf{c}_l^*(t)$ is the real coordinate.

## 2.3 Distributed Localization Estimation in 2D Space

Consider all follower agents located in convex hull, as shown in Fig. 1.6a. The objective of this section is to develop an efficient localization algorithm to estimate the real-time coordinates of the agents. In [1], Khan et al. propose a distributed iterative localization algorithm based on barycentric coordinates to guarantee the global convergence. Noted that Khan considers the localization problem of sensor nodes in static networks, only can ensure the localization estimation without time constraints. The real-time localization problem in time-varying dynamic networks should be considered in practical applications. Therefore, it is important to design a distributed localization estimation algorithm.

### 2.3.1 Distributed Localization with Convex Hull

Without loss of generality, the coordinate representation in two-dimensional space is considered. In order to estimate the accurate and real-time coordinate values of dynamic agents with repetitive characteristics, a distributed localization estimation algorithm based on iterative learning is proposed as follows

$$\hat{\mathbf{c}}_l^{i+1}(t) = \sum_{q \in \mathcal{N}_l} a_{l,q}^{i+1}(t)\hat{\mathbf{c}}_q^i(t) + \sum_{w \in \mathcal{N}_l} a_{l,w}^{i+1}(t)\mathbf{c}_w^*(t), \tag{2.1}$$

where $i$ represents the iteration number, $\mathcal{N}_l$ is the neighboring agent of $l$, $\mathbf{c}_w^*(t)$ and $\hat{\mathbf{c}}_q^i(t)$ represent the coordinate values of the neighboring leader agent and follower agent, respectively. $a_{l,q}^{i+1}(t)$ and $a_{l,w}^{i+1}(t)$ represent the calculated barycentric coordinate.

The complete process of localization estimation along the iteration axis is given in Algorithm 2.1.

**Algorithm 2.1**. Estimate the real coordinate of follower agent in convex hull

**for** $i = 1 : Iteration.$
**initialization** Give the initial iteration coordinate estimation value $\hat{\mathbf{c}}_l^i(t)$, $l \in \{m+1, \ldots, N\}$ of the follower agent randomly.
**Step 1.** Each agent transmits its own estimation value $\hat{\mathbf{c}}_l^i(t)$ to its neighbors.
**Step 2.** Measure the relative distance between agents, then calculate the barycentric coordinate $a_{l,q}^{i+1}(t)$ and$a_{l,w}^{i+1}(t)$.
**Step 3.** Update $\hat{\mathbf{c}}_l^{i+1}(t) = \sum_{q \in \mathcal{N}_l} a_{l,q}^{i+1}(t)\hat{\mathbf{c}}_q^i(t) + \sum_{w \in \mathcal{N}_l} a_{l,w}^{i+1}(t)\mathbf{c}_w^*(t)$
**Step 4.** Update $i = i + 1$ and return to **Step 1**.
**end**

**Remark 2.1** Algorithm 2.1 is based on iterative learning, since the estimated value $\hat{\mathbf{c}}_l^{i+1}(t)$ of $l$ in (2.1) is updated according to the estimation values of its neighboring agents at the $t$th sampling time of the $i$th iteration. The update of each agent is independent, and only local information is available from its neighboring agents. Therefore, Algorithm 2.1 is also distributed.

**Remark 2.2** Different from that of the distributed iterative localization algorithm proposed in [1], the coordinate estimation value in Algorithm 2.1 is updated along the iteration axis, which solves the real-time localization problem of dynamic multi-agent systems.

Consider all followers, one has a compact form

$$\hat{\mathbf{C}}_s^{i+1}(t) = \mathbf{P}^{i+1}(t)\hat{\mathbf{C}}_s^{i}(t) + \mathbf{B}^{i+1}(t)\mathbf{C}_a^*(t), \tag{2.2}$$

where $\mathbf{C}_a^*(t) = \left[\left(\mathbf{c}_1^*(t)\right)^{\mathrm{T}}, \ldots, \left(\mathbf{c}_m^*(t)\right)^{\mathrm{T}}\right]^{\mathrm{T}}$ represents the real coordinates of leaders. $\hat{\mathbf{C}}_s^{i}(t) = \left[\left(\hat{\mathbf{c}}_{m+1}^{i}(t)\right)^{\mathrm{T}}, \ldots, \left(\hat{\mathbf{c}}_{N}^{i}(t)\right)^{\mathrm{T}}\right]^{\mathrm{T}}$ represents the estimated coordinates of followers.

$\mathbf{P}^{i+1}(t) = \begin{bmatrix} 0 & a_{m+1,m+2}^{i+1}(t) & \cdots & a_{m+1,N}^{i+1}(t) \\ a_{m+2,m+1}^{i+1}(t) & 0 & \cdots & a_{m+2,N}^{i+1}(t) \\ \vdots & \vdots & \ddots & \vdots \\ a_{N,m+1}^{i+1}(t) & a_{N,m+2}^{i+1}(t) & \cdots & 0 \end{bmatrix} \in R^{(N-m)\times(N-m)}$ denotes the information transfer between follower agents.

$\mathbf{B}^{i+1}(t) = \begin{bmatrix} a_{m+1,1}^{i+1}(t) & a_{m+1,2}^{i+1}(t) & \cdots & a_{m+1,m}^{i+1}(t) \\ a_{m+2,1}^{i+1}(t) & a_{m+2,2}^{i+1}(t) & \cdots & a_{m+2,m}^{i+1}(t) \\ \vdots & \vdots & \ddots & \vdots \\ a_{N,1}^{i+1}(t) & a_{N,2}^{i+1}(t) & \cdots & a_{N,m}^{i+1}(t) \end{bmatrix} \in R^{(N-m)\times m}$ represents the information transfer between followers and leaders.

**Theorem 2.1** *Consider dynamic multi-agent systems with repetitive operation characteristics. By applying Algorithm 2.1 to solving the real-time localization problem, one can guarantee that* $\lim_{i\to\infty} \left\|\mathbf{E}_s^{i+1}(t)\right\|_\infty = 0$.

***Proof*** Define the estimation error as $\mathbf{E}_s^{i+1}(t) = \hat{\mathbf{C}}_s^{i+1}(t) - \mathbf{C}_s^*(t)$, where $\mathbf{C}_s^*(t)$ represents the real coordinates of all followers. Subtracting $\mathbf{C}_s^*(t)$ from both sides of (2.2), one has

$$\begin{aligned} &\hat{\mathbf{C}}_s^{i+1}(t) - \mathbf{C}_s^*(t) = \mathbf{P}^{i+1}(t)(\hat{\mathbf{C}}_s^{i}(t) - \mathbf{C}_s^*(t)) \\ &+\mathbf{P}^{i+1}(t)\mathbf{C}_s^*(t) - \mathbf{C}_s^*(t) + \mathbf{B}^{i+1}(t)\mathbf{C}_a^*(t). \end{aligned} \tag{2.3}$$

According to the definition of the estimation error, one has

$$\mathbf{E}_s^{i+1}(t) = \mathbf{P}^{i+1}(t)\mathbf{E}_s^{i}(t) + \mathbf{B}^{i+1}(t)\mathbf{C}_a^*(t) + \left(\mathbf{P}^{i+1}(t) - \mathbf{I}\right)\mathbf{C}_s^*(t). \tag{2.4}$$

Taking norms on both sides of (2.4), one has

$$\begin{aligned} &\left\|\mathbf{E}_s^{i+1}(t)\right\|_\infty \le \left\|\mathbf{P}^{i+1}(t)\right\|_\infty \left\|\mathbf{E}_s^{i}(t)\right\|_\infty + \\ &\left\|\mathbf{B}^{i+1}(t)\mathbf{C}_a^*(t) + \left(\mathbf{P}^{i+1}(t) - \mathbf{I}\right)\mathbf{C}_s^*(t)\right\|_\infty. \end{aligned} \tag{2.5}$$

The second term on the right-hand side of inequality (2.5) can be expanded as

$$\mathbf{B}^{i+1}(t)\mathbf{C}_a^*(t) + \left(\mathbf{P}^{i+1}(t) - \mathbf{I}\right)\mathbf{C}_s^*(t)$$
$$= \begin{bmatrix} a_{m+1,1}^{i+1}(t) & a_{m+1,2}^{i+1}(t) & \cdots & a_{m+1,m}^{i+1}(t) \\ a_{m+2,1}^{i+1}(t) & a_{m+2,2}^{i+1}(t) & \cdots & a_{m+2,m}^{i+1}(t) \\ \vdots & \vdots & \ddots & \vdots \\ a_{N,1}^{i+1}(t) & a_{N,2}^{i+1}(t) & \cdots & a_{N,m}^{i+1}(t) \end{bmatrix} \begin{bmatrix} \mathbf{c}_1^*(t) \\ \mathbf{c}_2^*(t) \\ \vdots \\ \mathbf{c}_m^*(t) \end{bmatrix} \tag{2.6}$$
$$+ \begin{bmatrix} -1 & a_{m+1,m+2}^{i+1}(t) & \cdots & a_{m+1,N}^{i+1}(t) \\ a_{m+2,m+1}^{i+1}(t) & -1 & \cdots & a_{m+2,N}^{i+1}(t) \\ \vdots & \vdots & \ddots & \vdots \\ a_{N,m+1}^{i+1}(t) & a_{N,m+2}^{i+1}(t) & \cdots & -1 \end{bmatrix} \begin{bmatrix} \mathbf{c}_{m+1}^*(t) \\ \mathbf{c}_{m+2}^*(t) \\ \vdots \\ \mathbf{c}_N^*(t) \end{bmatrix}.$$

By solving (2.6), one has

$$\begin{aligned} &a_{m+1,1}^{i+1}(t)\mathbf{c}_1^*(t) + \cdots + a_{m+1,m}^{i+1}(t)\mathbf{c}_m^*(t) + a_{m+1,m+2}^{i+1}(t) \\ &\quad \times \mathbf{c}_{m+2}^*(t) + \cdots + a_{m+1,N}^{i+1}(t)\mathbf{c}_N^*(t) - \mathbf{c}_{m+1}^*(t), \\ &\qquad \vdots \\ &a_{N,1}^{i+1}(t)\mathbf{c}_1^*(t) + \cdots + a_{N,m}^{i+1}(t)\mathbf{c}_m^*(t) + a_{N,m+1}^{i+1}(t) \\ &\quad \times \mathbf{c}_{m+1}^*(t) + \cdots + a_{N,N-1}^{i+1}(t)\mathbf{c}_{N-1}^*(t) - \mathbf{c}_N^*(t). \end{aligned} \tag{2.7}$$

Since $\mathbf{c}_w^*(t)$ represents the real coordinate, and through $\mathbf{c}_l(t) = a_{lh}(t)\mathbf{c}_h(t) + a_{lj}(t)\mathbf{c}_j(t) + a_{lk}(t)\mathbf{c}_k(t)$, one can know that when the real coordinates of the neighboring agents are known, the real coordinates of the agent to be located can be uniquely determined. It follows from $\mathbf{B}^{i+1}(t)\mathbf{C}_a^*(t) + \left(\mathbf{P}^{i+1}(t) - \mathbf{I}\right)\mathbf{C}_s^*(t)$ that

$$\begin{aligned} &a_{m+1,1}^{i+1}(t)\mathbf{c}_1^*(t) + \cdots + a_{m+1,m}^{i+1}(t)\mathbf{c}_m^*(t) + a_{m+1,m+2}^{i+1}(t) \\ &\quad \mathbf{c}_{m+2}^*(t) + \cdots + a_{m+1,N}^{i+1}(t)\mathbf{c}_N^*(t) = \mathbf{c}_{m+1}^*(t), \\ &\qquad \vdots \\ &a_{N,1}^{i+1}(t)\mathbf{c}_1^*(t) + \cdots + a_{N,m}^{i+1}(t)\mathbf{c}_m^*(t) + a_{N,m+1}^{i+1}(t)\mathbf{c}_{m+1}^*(t) \\ &\quad + \cdots + a_{N,N-1}^{i+1}(t)\mathbf{c}_{N-1}^*(t) = \mathbf{c}_N^*(t). \end{aligned} \tag{2.8}$$

Equation (2.8) means

$$\left\|\mathbf{B}^{i+1}(t)\mathbf{C}_a^*(t) + \left(\mathbf{P}^{i+1}(t) - \mathbf{I}\right)\mathbf{C}_s^*(t)\right\|_\infty = \mathbf{0}. \tag{2.9}$$

Since the element in matrix $\mathbf{P}^{i+1}(t)$ represents the barycentric coordinate between follower agents, it can be guaranteed that the sum of each row of matrix $\mathbf{P}^{i+1}(t)$ is less than 1. Then, according to the property of infinite norm, one has $0 < \left\|\mathbf{P}^{i+1}(t)\right\|_\infty = \sigma_1 < 1$.

Based on the analysis from (2.3) to (2.8), the following result can be obtained.

$$\begin{aligned}\left\|\mathbf{E}_s^{i+1}(t)\right\|_\infty &\le \sigma_1\left\|\mathbf{E}_s^{i}(t)\right\|_\infty \le \sigma_1^2\left\|\mathbf{E}_s^{i-1}(t)\right\|_\infty\\ &\le \cdots \le \sigma_1^{i+1}\left\|\mathbf{E}_s^{0}(t)\right\|_\infty.\end{aligned} \tag{2.10}$$

Due to the fact that $0 < \sigma_1 < 1$ and the initial estimation error $\left\|\mathbf{E}_s^{0}(t)\right\|_\infty$ is bounded, (2.10) implies $\lim_{i\to\infty}\left\|\mathbf{E}_s^{i+1}(t)\right\|_\infty = 0$. Thus, the proof of Theorem 2.1 is complete.

### 2.3.2 *Distributed Localization with Arbitrary Deployment*

This section discusses the case shown in Fig. 1.6b, i.e., agent $l$ is outside the convex hull formed by its neighboring agents. In such case, the method presented in Sect. 2.3.1 is not applicable because the sum of the calculated barycentric coordinates is greater than 1. The objective of this section is to develop a distributed localization estimation algorithm, which addresses the real-time localization problem of dynamic multi-agent systems with arbitrary deployment.

First, how to determine the sign of the barycentric coordinates need to be considered.

Consider agent $l$ and its three neighbors $h$, $j$ and $k$. As shown in Fig. 2.1, when agent $l$ is deployed arbitrarily, there will be six different situations $\{1, 2, 3, 4, 5, 6\}$. Moreover, the calculated barycentric coordinates cannot satisfy the condition $a_{l,h}(t) + a_{l,j}(t) + a_{l,k}(t) = 1$. Therefore, these six different deployment situations should be analyzed to ensure the accurate localization. Here, a signed time-varying scalar coefficient $\gamma(t)$, whose value is $+1$ or $-1$. The specific algorithm is given as:

$$\gamma_{l,h}(t)a_{l,h}(t) + \gamma_{l,j}(t)a_{l,j}(t) + \gamma_{l,k}(t)a_{l,k}(t) = 1. \tag{2.11}$$

The barycentric coordinates of agent $l$ satisfy $\mathbf{c}_l(t) = \gamma_{l,h}(t)a_{l,h}(t)\mathbf{c}_h(t) + \gamma_{l,j}(t)a_{l,j}(t)\mathbf{c}_j(t) + \gamma_{l,k}(t)a_{l,k}(t)\mathbf{c}_k(t)$.

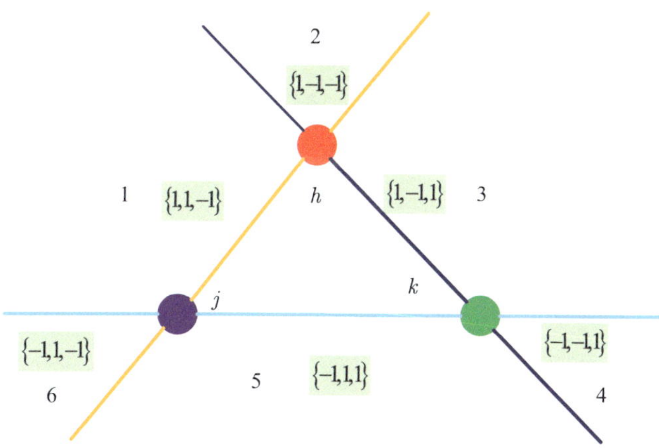


**Fig. 2.1** Six possible deployment situations

It follows from (2.11) that when agent $l$ is located in the area $\{1, 2, 3, 4, 5, 6\}$, the sign coefficients $\{\gamma_{l,h}(t), \gamma_{l,j}(t), \gamma_{l,k}(t)\}$ of the barycentric coordinates are $\{1, 1, -1\}$, $\{1, -1, -1\}$, $\{1, -1, 1\}$, $\{-1, -1, 1\}$, $\{-1, 1, 1\}$ and $\{-1, 1, -1\}$.

Sometimes, the solution of (2.11) is not unique. Next, two specific analysis of these cases are given.

**Case 1** Agent $l$ is located in the junction of two regions.

When agent $l$ is located at the junction of two regions, one of its barycentric coordinates is equal to zero, and then (2.11) will have two different solutions. Therefore, in order to make the localization algorithm work normally and achieve accurate localization, a distinguishing mechanism is proposed as follows.

For example, as shown in Fig. 2.2, when agent $l$ is located at point $l'$, the barycentric coordinates $a_{l,h}(t) = 0$. Two different solutions $\{1, 1, -1\}$ and $\{-1, 1, -1\}$ are obtained by solving (2.11). These two solutions are feasible and will not have adverse effects on the accurate localization. To avoid unnecessary confusion, when $a_{l,h}(t) = 0$, the sign coefficient is set as $\gamma_{l,h}(t) = 1$, and then the values of the remaining two sign coefficients can be uniquely determined by (2.11). Similarly, when $a_{l,j}(t) = 0$ and $a_{l,k}(t) = 0$, $\gamma_{l,j}(t) = 1$ and $\gamma_{l,k}(t) = 1$ can be set respectively to make the solution of (2.10) unique.

**Case 2** Three different solutions for (2.11)

When $a_{l,h}(t) = 1$ and $a_{l,j}(t) = a_{l,k}(t)$, it can be found that there are three different solutions for (2.11). Note that the position of agent $l$ and its neighboring agents are not coincident, so the situation $a_{l,j}(t) = a_{l,k}(t) = 0$ does not exist. Similar to the Case 1, in order to achieve fast and accurate localization, it is necessary to design a distinguishing method to determine the value of the sign coefficient $\{\gamma_{l,h}(t), \gamma_{l,j}(t), \gamma_{l,k}(t)\}$ of $l$.

Before designing the distinguishing method, several possible positions of agent $l$ need be introduced. As shown in Fig. 2.3, $r$ represents a dotted line parallel to $jk$. $l'$,

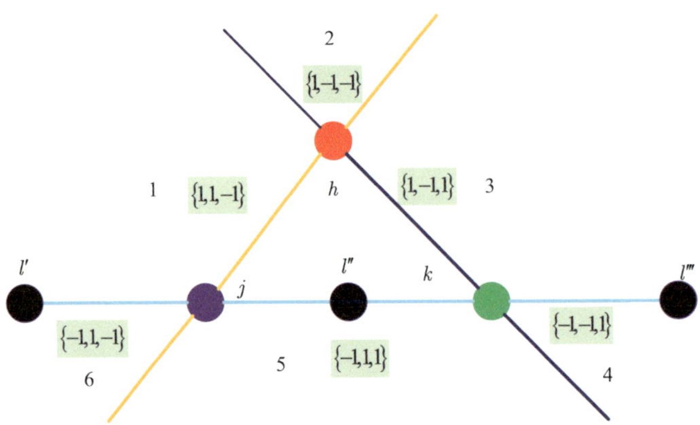


**Fig. 2.2** Agent located in the junction of two regions

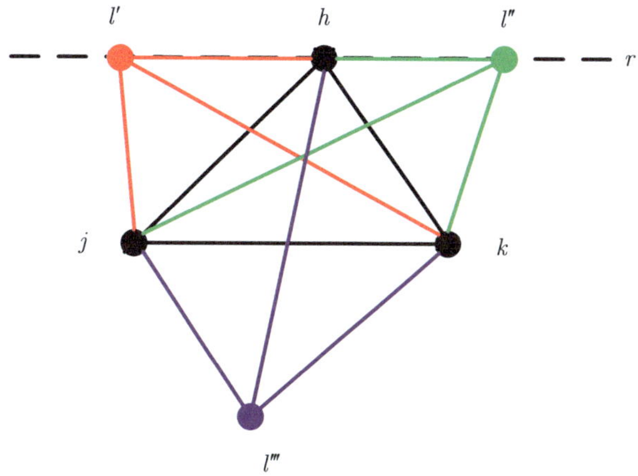


**Fig. 2.3** Three different solutions for (10)

$l''$, and $l'''$ are three possible positions of agent $l$ at the $t$th sampling time, where $jl'''$ is parallel to $hk$ and $jh$ is parallel to $kl'''$.

**Lemma 2.1** *The barycentric coordinates satisfy* $a_{l,h}(t) =$ *and* $a_{l,j}(t) = a_{l,k}(t)$, *when agent* $l$ *is located at any point on the dotted line* $r$ *or* $l'''$ *at the* $t$th *sampling time.*

***Proof*** Here, $l'$ and $l'''$ in Fig. 2.3 are used to prove the Lemma. When agent $l$ moves to point $l'$, the heights of triangles $\Delta ljk$ and $\Delta hjk$ are equal since the dotted line $r$ is parallel to $jk$. The two triangles use the same bottom $jk$. It can be obtained that $S_{\Delta ljk}(\text{t}) = S_{\Delta hjk}(t)$, $a_{l,h}(t) = 1$. In the same way, it can be proved that $S_{\Delta lkh} = S_{\Delta lhj}$, $a_{l,j}(t) = a_{l,k}(t)$.

When agent $l$ moves to point $l'''$, since agents $h$, $j$, $k$ and $l$ form a parallelogram, according to the properties of parallelogram, one can get that $S_{\Delta ljk}(\text{t}) = S_{\Delta hjk}(t)$, $a_{l,h}(t) = 1$ and $S_{\Delta lkh} = S_{\Delta lhj}$, $a_{l,j}(t) = a_{l,k}(t)$.

Thus, the proof of the Lemma 2.1 is complete.

According to the analysis of the Case 2, the specific distinguishing algorithm is given as follows:

$$\begin{aligned} \textbf{if} \quad & d_{hj}(t) = d_{lk}(t), d_{jl}(t) = d_{hk}(t) \quad and \\ & d_{hl}^2(t) + d_{jk}^2(t) = 2d_{hj}^2(t) + 2d_{hk}^2(t), \\ \textbf{then} \quad & \left\{\gamma_{l,h}(t), \gamma_{l,j}(t), \gamma_{l,k}(t)\right\} = \{-1, 1, 1\}. \end{aligned} \tag{2.12}$$

$$\begin{aligned} \textbf{if} \quad & d_{jl}^2(t) < d_{hj}^2(t) + d_{hl}^2(t), \\ \textbf{then} \quad & \left\{\gamma_{l,h}(t), \gamma_{l,j}(t), \gamma_{l,k}(t)\right\} = \{1, 1, -1\}. \end{aligned} \tag{2.13}$$

$$\begin{aligned} \textbf{if} \quad & d_{jl}^2(t) > d_{hj}^2(t) + d_{hl}^2(t), \\ \textbf{then} \quad & \left\{\gamma_{l,h}(t), \gamma_{l,j}(t), \gamma_{l,k}(t)\right\} = \{1, -1, 1\}. \end{aligned} \tag{2.14}$$

Equations (2.12)–(2.14) are derived from the cosine law of triangles, and the details are omitted here.

Based on the analysis from (2.12) to (2.14), a detailed and complete process to determine the sign coefficients of barycentric coordinates is given in Algorithm 2.2.

**Algorithm 2.2**. Determine the sign coefficients of barycentric coordinates

**for** $i = 1 : Iteration.$
**input** $a_{l,h}(t), a_{l,j}(t), a_{l,k}(t), d_{hl}(t), d_{hj}(t), d_{hk}(t), d_{jl}(t), d_{jk}(t), d_{lk}(t)$.
**solve** equation (2.11).
**if** the solution is unique,
**then** skip to **output**.
**else if** one of barycentric coordinates $a$ is equal to 0,
**then** set the corresponding sign coefficient $\gamma$ equal to 1 and go back to **solve**.
**else if** $d_{hl}^2(t) + d_{jk}^2(t) = 2d_{hj}^2(t) + 2d_{hk}^2(t)$, $d_{hj}(t) = d_{lk}(t)$ and $d_{jl}(t) = d_{hk}(t)$,
**then** $\{\gamma_{l,h}(t), \gamma_{l,j}(t), \gamma_{l,k}(t)\} = \{-1, 1, 1\}$.
**else if** $d_{jl}^2(t) < d_{hj}^2(t) + d_{hl}^2(t)$ ,
**then** $\{\gamma_{l,h}(t), \gamma_{l,j}(t), \gamma_{l,k}(t)\} = \{1, 1, -1\}$.
**else if** $d_{jl}^2(t) > d_{hj}^2(t) + d_{hl}^2(t)$,
**then** $\{\gamma_{l,h}(t), \gamma_{l,j}(t), \gamma_{l,k}(t)\} = \{1, -1, 1\}$.
**output** $\{\gamma_{l,h}(t), \gamma_{l,j}(t), \gamma_{l,k}(t)\}$.
**end**

After determining the sign coefficients of the barycentric coordinates, a distributed localization estimation algorithm with sign coefficients is proposed

$$\hat{\mathbf{c}}_l^{i+1}(t) = \sum_{q\in N_l} \gamma_{l,q}^{i+1}(t)a_{l,q}^{i+1}(t)\hat{\mathbf{c}}_q^i(t) + \sum_{w\in N_l} \gamma_{l,w}^{i+1}(t)a_{l,w}^{i+1}(t)\mathbf{c}_w^*(t). \tag{2.15}$$

The difference between (2.15) and (2.1) is that the sign coefficients $\gamma_{l,q}^{i+1}(t)$ and $\gamma_{l,w}^{i+1}(t)$ are added to ensure $\gamma_{l,h}(t)a_{l,h}(t) + \gamma_{l,j}(t)a_{l,j}(t) + \gamma_{l,k}(t)a_{l,k}(t) = 1$.

For all $(N - m)$ follower agents, the compact form of (2.15) is as follows:

$$\hat{\mathbf{C}}_s^{i+1}(t) = \mathbf{G}^{i+1}(t)\hat{\mathbf{C}}_s^i(t) + \Phi^{i+1}(t)\mathbf{C}_a^*(t), \tag{2.16}$$

where $\Phi^{i+1}(t) = \begin{bmatrix} \gamma_{m+1,1}^{i+1}(t)a_{m+1,1}^{i+1}(t) & \cdots & \gamma_{m+1,m}^{i+1}(t)a_{m+1,m}^{i+1}(t) \\ \vdots & \ddots & \vdots \\ \gamma_{N,1}^{i+1}(t)a_{N,1}^{i+1}(t) & \cdots & \gamma_{N,m}^{i+1}(t)a_{N,m}^{i+1}(t) \end{bmatrix}$,

$$\mathbf{G}^{i+1}(t) = \begin{bmatrix} 0 & \begin{matrix}\gamma_{m+1,m+2}^{i+1}(t)\\ \times a_{m+1,m+2}^{i+1}(t)\end{matrix} & \cdots & \begin{matrix}\gamma_{m+1,N}^{i+1}(t)\\ \times a_{m+1,N}^{i+1}(t)\end{matrix} \\ \begin{matrix}\gamma_{m+2,m+1}^{i+1}(t)\\ \times a_{m+2,m+1}^{i+1}(t)\end{matrix} & 0 & \cdots & \begin{matrix}\gamma_{m+2,N}^{i+1}(t)\\ \times a_{m+2,N}^{i+1}(t)\end{matrix} \\ \vdots & \vdots & \ddots & \vdots \\ \begin{matrix}\gamma_{N,m+1}^{i+1}(t)\\ \times a_{N,m+1}^{i+1}(t)\end{matrix} & \begin{matrix}\gamma_{N,m+2}^{i+1}(t)\\ \times a_{N,m+2}^{i+1}(t)\end{matrix} & \cdots & 0 \end{bmatrix}.$$

According to the analysis in Sect. 2.3.1, note that matrix $\mathbf{G}^{i+1}(t)$ may be unstable when follower agents are deployed arbitrarily, so (2.16) may not converge.

Inspired by Richardson iteration method [11], an improved distributed localization estimation algorithm based on iterative learning is proposed.

$$\hat{\mathbf{C}}_s^{i+1}(t) = \left(\mathbf{I} - \delta\big(\mathbf{Q}^{i+1}(t)\big)^{\mathrm{T}}\mathbf{Q}^{i+1}(t)\right)\hat{\mathbf{C}}_s^{i}(t) + \delta\big(\mathbf{Q}^{i+1}(t)\big)^{\mathrm{T}}\Phi^{i+1}(t)\mathbf{C}_a^*(t), \quad (2.17)$$

where $\delta$ is a scalar coefficient, $\mathbf{Q}^{i+1}(t) = \mathbf{I} - \mathbf{G}^{i+1}(t)$, $\mathbf{I} - \mathbf{G}^{i+1}(t)$ is nonsingular and $\big(\mathbf{Q}^{i+1}(t)\big)^{\mathrm{T}}\mathbf{Q}^{i+1}(t)$ is positive definite.

The process of the distributed localization estimation algorithm for dynamic multi-agent systems with arbitrary deployment is presented in Algorithm 2.3.

---

**Algorithm 2.3**. Estimate the real coordinates of the follower agents with arbitrary deployment

---

**for** $i = 1 : Iteration.$

**initialization** Give the initial iteration coordinate estimation value $\hat{\mathbf{c}}_l^i(t)$, $l \in \{m+1, \ldots, N\}$ of the follower agent randomly and select the scalar coefficient $\delta$ properly.

**Step 1.** Each agent transmits its own estimation value $\hat{\mathbf{c}}_l^i(t)$ to its neighbors.

**Step 2.** Measure the relative distance between agents and calculate the barycentric coordinate $a_{l,q}^{i+1}(t)$ and $a_{l,w}^{i+1}(t)$.

**Step 3.** Calculate the sign coefficients $\gamma_{l,q}^{i+1}(t)$ and $\gamma_{l,w}^{i+1}(t)$ by using Algorithm 2.2.

**Step 4.** Update

$\hat{\mathbf{C}}_s^{i+1}(t) = \left(\mathbf{I} - \delta\big(\mathbf{Q}^{i+1}(t)\big)^{\mathrm{T}}\mathbf{Q}^{i+1}(t)\right)\hat{\mathbf{C}}_s^{i}(t) + \delta\big(\mathbf{Q}^{i+1}(t)\big)^{\mathrm{T}}\Phi^{i+1}(t)\mathbf{C}_a^*(t).$

**Step 5.** Update $i = i + 1$ and return to **Step 1**.

**end**

---

Consider all the followers, the convergence result on the localization estimation error is given below.

**Theorem 2.2** *Consider dynamic multi-agent systems with repetitive operation characteristics under arbitrary deployment. Apply the distributed estimation Algorithm 2.3 to solving the real-time localization problem. One can guarantee that* $\lim_{i\to\infty} \left\|\mathbf{E}_s^{i+1}(t)\right\|_2 = 0$.

***Proof*** Similar to Sect. 2.3.1, the estimation errors $\mathbf{E}_s^{i+1}(t) = \hat{\mathbf{C}}_s^{i+1}(t) - \mathbf{C}_s^*(t)$ is defined. Subtracting $\mathbf{C}_s^*(t)$ from both sides of (2.17) yields

$$\begin{aligned}\mathbf{E}_s^{i+1}(t) = &\left(\mathbf{I} - \delta\big(\mathbf{Q}^{i+1}(t)\big)^{\mathrm{T}}\mathbf{Q}^{i+1}(t)\right)\mathbf{E}_s^{i}(t) + \\ &\delta\big(\mathbf{Q}^{i+1}(t)\big)^{\mathrm{T}}\left(\Phi^{i+1}(t)\mathbf{C}_a^*(t) - \mathbf{Q}^{i+1}(t)\mathbf{C}_s^*(t)\right).\end{aligned} \quad (2.18)$$

Since $\mathbf{c}^*(t)$ represents the real coordinate, and through $\mathbf{c}_l(t) = \gamma_{l,h}(t)a_{l,h}(t)\mathbf{c}_h(t) + \gamma_{l,j}(t)a_{l,j}(t)\mathbf{c}_j(t) + \gamma_{l,k}(t)a_{l,k}(t)\mathbf{c}_k(t)$, one can know that when the real coordinates of the neighboring agents are known, the real coordinates of the agent to be located can be uniquely determined. It follows from $\Phi^{i+1}(t)\mathbf{C}_a^*(t) - \mathbf{Q}^{i+1}(t)\mathbf{C}_s^*(t)$ that

$$
\begin{aligned}
&\gamma_{m+1,1}^{i+1}(t)a_{m+1,1}^{i+1}(t)\mathbf{c}_1^*(t) + \cdots + \gamma_{m+1,m}^{i+1}(t)a_{m+1,m}^{i+1}(t) \\
&\mathbf{c}_m^*(t) + \gamma_{m+1,m+2}^{i+1}(t)a_{m+1,m+2}^{i+1}(t)\mathbf{c}_{m+2}^*(t) + \cdots + \\
&\gamma_{m+1,N}^{i+1}(t)a_{m+1,N}^{i+1}(t)\mathbf{c}_N^*(t) = \mathbf{c}_{m+1}^*(t), \\
&\vdots \\
&\gamma_{N,1}^{i+1}(t)a_{N,1}^{i+1}(t)\mathbf{c}_1^*(t) + \cdots + \gamma_{N,m}^{i+1}(t)a_{N,m}^{i+1}(t)\mathbf{c}_m^*(t) \\
&+\gamma_{N,m+1}^{i+1}(t)a_{N,m+1}^{i+1}(t)\mathbf{c}_{m+1}^*(t) + \cdots + \\
&\gamma_{N,N-1}^{i+1}(t)a_{N,N-1}^{i+1}(t)\mathbf{c}_{N-1}^*(t) = \mathbf{c}_N^*(t),
\end{aligned}
\tag{2.19}
$$

which means $\delta\big(\mathbf{Q}^{i+1}(t)\big)^{\mathrm{T}}\big(\Phi^{i+1}(t)\mathbf{C}_a^*(t) - \mathbf{Q}^{i+1}(t)\mathbf{C}_s^*(t)\big) = \mathbf{0}$.

It follows from (2.18) that

$$
\mathbf{E}_s^{i+1}(t) = \Big(\mathbf{I} - \delta\big(\mathbf{Q}^{i+1}(t)\big)^{\mathrm{T}}\mathbf{Q}^{i+1}(t)\Big)\mathbf{E}_s^i(t). \tag{2.20}
$$

The necessary and sufficient condition for the convergence of Algorithm 2.3 is that the spectral radius of matrix $\Big(\mathbf{I} - \delta\big(\mathbf{Q}^{i+1}(t)\big)^{\mathrm{T}}\mathbf{Q}^{i+1}(t)\Big)$ is less than 1. Select $\delta$ on the open interval $0 < \delta < 2/\lambda_{\max}$, where $\lambda_{\max}$ is the maximum eigenvalue of $\big(\mathbf{Q}^{i+1}(t)\big)^{\mathrm{T}}\mathbf{Q}^{i+1}(t)$. Then, the matrix $\Big(\mathbf{I} - \delta\big(\mathbf{Q}^{i+1}(t)\big)^{\mathrm{T}}\mathbf{Q}^{i+1}(t)\Big)$ in (2.20) is stable and Algorithm 2.3 is convergent. Furthermore, if $\delta = 2/(\lambda_{\min} + \lambda_{\max})$, the fastest convergence rate can be obtained, where $\lambda_{\min}$ is the minimum eigenvalue of $\big(\mathbf{Q}^{i+1}(t)\big)^{\mathrm{T}}\mathbf{Q}^{i+1}(t)$. Refer to [12] for details.

Taking norms on (2.20), one has

$$
\big\|\mathbf{E}_s^{i+1}(t)\big\|_2 \le \Big\|\mathbf{I} - \delta\big(\mathbf{Q}^{i+1}(t)\big)^{\mathrm{T}}\mathbf{Q}^{i+1}(t)\Big\|_2 \big\|\mathbf{E}_s^i(t)\big\|_2. \tag{2.21}
$$

Let $\sigma_2 = \Big\|\mathbf{I} - \delta\big(\mathbf{Q}^{i+1}(t)\big)^{\mathrm{T}}\mathbf{Q}^{i+1}(t)\Big\|_2$, and $0 < \sigma_2 < 1$ can be guaranteed by selecting $\delta$ properly. Then, the following results can be obtained

$$
\big\|\mathbf{E}_s^{i+1}(t)\big\|_2 \le \sigma_2\big\|\mathbf{E}_s^i(t)\big\|_2 \le \sigma_2^2\big\|\mathbf{E}_s^{i-1}(t)\big\|_2 \le \cdots \le \sigma_2^{i+1}\big\|\mathbf{E}_s^0(t)\big\|_2. \tag{2.22}
$$

From (2.22), one can get that $\lim_{i\to\infty}\big\|\mathbf{E}_s^{i+1}(t)\big\|_2 = 0$. Thus, the proof of Theorem 2.2 is complete.

### 2.3.3 Corollary

Under some specific topological structures of multi-agent systems, the estimator (2.17) can be simplified. Next, some discussion and analysis for this case are given.

For asymmetric non-singular matrix $\mathbf{Q}^{i+1}(t)$, if it can be diagonalized and its eigenvalues are all positive, a simplified localization estimation algorithm based on Richardson iteration method can be designed as

$$\hat{\mathbf{C}}_s^{i+1}(t) = \left(\mathbf{I} - \delta\mathbf{Q}^{i+1}(t)\right)\hat{\mathbf{C}}_s^{i}(t) + \delta\Phi^{i+1}(t)\mathbf{C}_a^{i}(t). \tag{2.23}$$

For convenience, the intermediate analysis process is omitted. Similar to the analysis in Sect. 2.3.2, one has

$$\mathbf{E}_s^{i+1}(t) = \left(\mathbf{I} - \delta\mathbf{Q}^{i+1}(t)\right)\mathbf{E}_s^{i}(t). \tag{2.24}$$

For asymmetric matrix $\mathbf{Q}^{i+1}(t)$, one can find a similar nonsingular transformation matrix $\mathbf{X}^{i+1}(t)$ such that

$$\left(\mathbf{X}^{i+1}(t)\right)^{-1}\mathbf{Q}^{i+1}(t)\mathbf{X}^{i+1}(t) = \Lambda = diag(\lambda_1^{i+1}(t), \ldots, \lambda_{N-m}^{i+1}(t)), \tag{2.25}$$

where $\left\{\lambda_1^{i+1}(t), \ldots, \lambda_{N-m}^{i+1}(t)\right\}$ are eigenvalues of $\mathbf{Q}^{i+1}(t)$.

Based on similar transformation matrix $\mathbf{X}^{i+1}(t)$ and following from (2.25), one can obtain that

$$\mathbf{E}_s^{i+1}(t) = \mathbf{X}^{i+1}(t)diag(1 - \delta\lambda_1^{i+1}(t), \ldots, 1 - \delta\lambda_{N-m}^{i+1}(t))\left(\mathbf{X}^{i+1}(t)\right)^{-1}\mathbf{E}_s^{i}(t). \tag{2.26}$$

Note that the necessary and sufficient condition for convergence of (2.24) is that the modulus of all eigenvalues of $\left(\mathbf{I} - \delta\mathbf{Q}^{i+1}(t)\right)$ is less than 1, i.e., $\rho\left(\mathbf{I} - \delta\mathbf{Q}^{i+1}(t)\right) < 1$, where $\rho$ is the spectral radius.

Define $\rho\left(\mathbf{I} - \delta\mathbf{Q}^{i+1}(t)\right) = \max\left\{\left|1 - \delta\lambda_1^{i+1}(t)\right|, \ldots, \left|1 - \delta\lambda_{N-m}^{i+1}(t)\right|\right\}$. Select $0 < \delta < 2/\lambda_{\max}$, where $\lambda_{\max} = \max\left\{\left|\lambda_1^{i+1}(t)\right|, \ldots, \left|\lambda_{N-m}^{i+1}(t)\right|\right\}$. One has $\left|1 - \delta\lambda_\kappa^{i+1}(t)\right| < 1, \kappa = 1, \ldots, N - m$. Then, $\rho\left(\mathbf{I} - \delta\mathbf{Q}^{i+1}(t)\right) < 1$ is obtained.

The rest of the proof is similar to Theorem 2.2 and is omitted here.

## 2.4 Distributed Localization Estimation in 3-D Space

This section will discuss the three-dimensional (3-D) distributed localization problem for the multi-agent systems. The objective of this section is to develop a distributed localization estimation algorithm, which solves the real-time localization problem of dynamic multi-agent systems with arbitrary deployment.

It should be noted that in Fig. 1.7a, the agent to be estimated is always located in the tetrahedron formed by its neighbouring agents, which limits the application of the localization algorithm in practical systems. In fact, as shown in Fig. 1.7b, the neighboring agents of agent $l$ form a tetrahedron, and agent $l$ is deployed arbitrarily. The position of agent $l$ and its neighboring agents $h$, $j$, $k$ and $n$ does not overlap. When the agent $l$ is deployed arbitrarily in 3D space, condition $a_{l,n}(t) + a_{l,h}(t) +$

**Fig. 2.4** Fifteen possible deployment situations

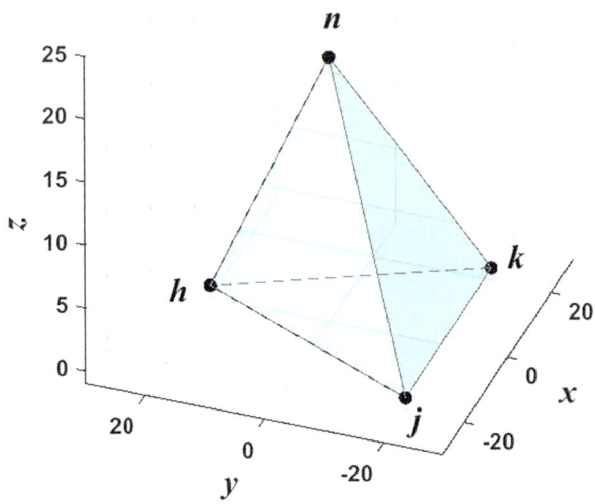


$a_{l,j}(t) + a_{l,k}(t) = 1$ is usually not satisfied and the coordinates of the agent $l$ will not be estimated using algorithm (2.1).

Similar to Sect. 2.3.2, the signed time-varying scalar coefficient $\gamma(t)$ is introduced to guarantee the localizability of the multi-agent systems. An algorithm is given below

$$\gamma_{l,n}(t)a_{l,n}(t) + \gamma_{l,h}(t)a_{l,h}(t) + \gamma_{l,j}(t)a_{l,j}(t) + \gamma_{l,k}(t)a_{l,k}(t) = 1. \tag{2.27}$$

Thus, the barycentric coordinates of agent $l$ satisfy

$$\begin{aligned}\mathbf{c}_l(t) = \gamma_{l,n}(t)a_{l,n}(t)\mathbf{c}_n(t) + \gamma_{l,h}(t)a_{l,h}(t)\mathbf{c}_h(t) + \\ \gamma_{l,j}(t) \times a_{l,j}(t)\mathbf{c}_j(t) + \gamma_{l,k}(t)a_{l,k}(t)\mathbf{c}_k(t).\end{aligned} \tag{2.28}$$

As shown in Fig. 2.4, the 3-D space will be divided into 15 different subspaces when the four planes of the tetrahedron extend infinitely. Among which, four vertices $l, n, h, j$ and $k$ correspond to four different subspaces, six edges $nh, nj, nk, hj, hk$ and $jk$ correspond to six different subspaces, four planes $nhk$, $nhj$, $njk$ and $hjk$ correspond to four different subspaces, and the interior of the tetrahedron $nhjk$ also corresponds to one subspace. The sign coefficients of the barycentric coordinates vary with different subspaces.

By solving (2.27), $\left(\gamma_{l,n}(t), \gamma_{l,h}(t), \gamma_{l,j}(t), \gamma_{l,k}(t)\right)$ are given as follows $\{(1,-1,-1,-1)\,|l \in n\_sub\}, \{(-1,1,-1,-1)\,|l \in h\_sub\}, \{(-1,-1,1,-1)\,|l \in j\_sub\}$, $\{(-1,-1,-1,1)\,|l \in k\_sub\}$, $\{(1,1,-1,1)\,|l \in nhk\_sub\}$, $\{(1,1,1,-1)\,|l \in nhj\_sub\}$, $\{(1,-1,1,1)\,|l \in njk\_sub\}$, $\{(-1,1,1,1)\,|l \in hjk\_sub\}$, $\{(1,1,-1,-1)\,|l \in nh\_sub\}$,

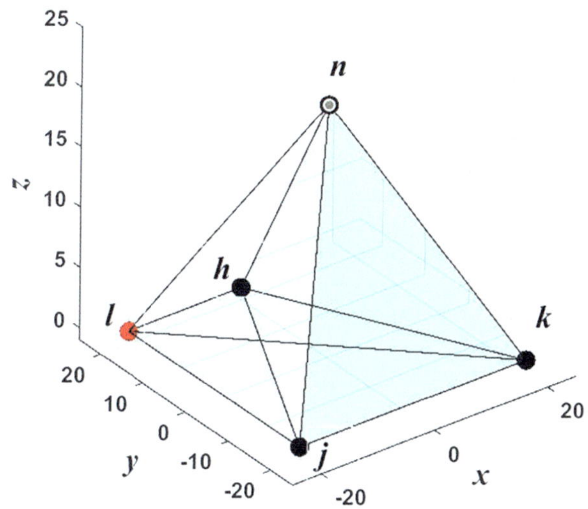


**Fig. 2.5** Agent $l$ is located on the extension surface of a plane

$\{(1,-1,1,-1)\,|l \in nj\_sub\}$, $\{(1,-1,-1,1)\,|l \in nk\_sub\}$, $\{(-1,1,1,-1)\,|l \in hj\_sub\}$, $\{(-1,1,-1,1)\,|l \in hk\_sub\}$, $\{(-1,-1,1,1)\,|l \in jk\_sub\}$, $\{(1,1,1,1)\,|l \in nhjk\_sub\}$.

Some abbreviations are introduced for the purpose of saving space, for example, $n\_sub$ represents the subspace correspond to the vertex $n$.

Generally, $\left(\gamma_{l,n}(t), \gamma_{l,h}(t), \gamma_{l,j}(t), \gamma_{l,k}(t)\right)$ can be uniquely determined by (2.27). However, the solution of (2.27) is not unique when agent $l$ is located in some special regions. The detailed analysis of these special cases is given as follows.

**Case 1** Agent $l$ is located on the plane.

As shown in Fig. 2.5, when agent $l$ is located on the extension surface of the $jhk$ plane, the barycentric coordinates $a_{l,n}(t)$ is equal to zero. and then (2.27) will have two different solutions $\{-1, 1, 1, 1\}$ and $\{-1, 1, -1, 1\}$. In order to avoid unnecessary confusion, the sign coefficient of the $a_{l,n}(t)$ is set as $\gamma_{l,n}(t) = 1$. Similarly, as shown in Fig. 2.6, when agent $l$ is located on the surface of the $nhj$ plane, one can set as $\gamma_{l,k}(t) = 1$ to make the solution of (2.27) is unique.

**Case 2** Agent $l$ is located in the junction of two planes.

As shown in Fig. 2.7, when agent $l$ is located on the junction line of the two planes, (2.27) will have four different solutions. For example, the barycentric coordinates $a_{l,h}(t) = 0$ and $a_{l,k}(t) = 0$ when agent $l$ is located at $l_1$. Then, there are four different solutions $(1, -1, -1, -1)$, $(1, 1, -1, 1)$, $(1, 1, -1, -1)$ and $(1, -1, -1, 1)$, which can make (2.27) hold. Similarly, there are also four different solutions when agent $l$

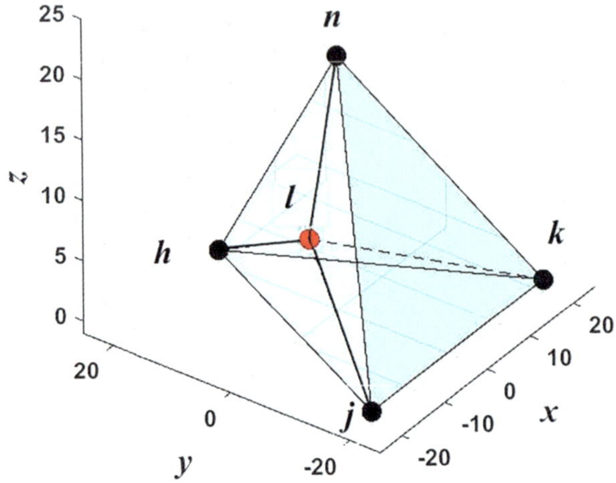


**Fig. 2.6** Agent $l$ is located on the surface of a plane

**Fig. 2.7** The agent is located in the junction of two planes

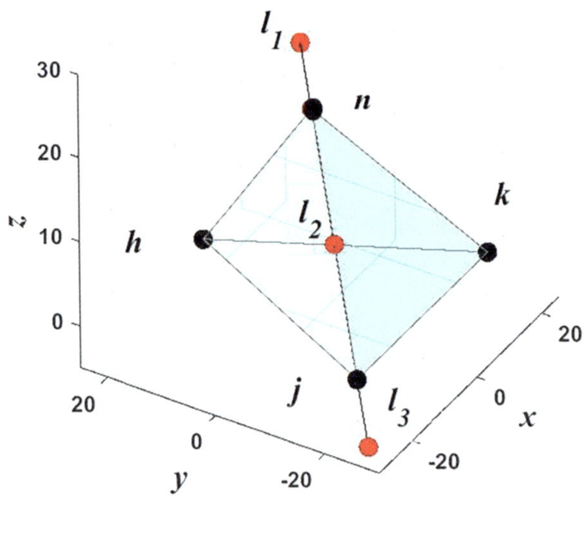


is located at $l_2$ or $l_3$. In this case, let $\gamma_{l,h}(t) = 1$ and $\gamma_{l,k}(t) = 1$ to avoid unnecessary confusion.

In order to determine the value of sign coefficients $\left(\gamma_{l,n}(t), \gamma_{l,h}(t), \gamma_{l,j}(t), \gamma_{l,k}(t)\right)$ in Case 2, a discrimination mechanism is given under the premise of barycentric coordinates $a_{l,h}(t) = 0$ and $a_{l,k}(t) = 0$.

If $d_{jn}(t) + d_{nl}(t) = d_{jl}(t)$, i.e., agent $l$ is located at $l_1$, then $\left(\gamma_{l,n}(t), \chi_{l,h}(t), \gamma_{l,j}(t), \gamma_{l,k}(t)\right) = (1, 1, -1, 1)$,

If $d_{jl}(t) + d_{nl}(t) = d_{jn}(t)$, i.e., agent $l$ is located at $l_2$, then $\left(\gamma_{l,n}(t), \gamma_{l,h}(t), \gamma_{l,j}(t), \gamma_{l,k}(t)\right) = (1, 1, 1, 1)$,

If $d_{jl}(t) + d_{jn}(t) = d_{nl}(t)$, i.e., agent $l$ is located at $l_3$, then $\left(\gamma_{l,n}(t), \chi_{l,h}(t), \gamma_{l,j}(t), \gamma_{l,k}(t)\right) = (-1, 1, 1, 1)$.

Based on the above analysis, the concise and complete process to determine the sign coefficients of the barycentric coordinates is provided in Algorithm 2.4.

---

**Algorithm 2.4**. Determine the sign coefficients of barycentric coordinates

---

**input** $a_{l,n}(t)$, $a_{l,h}(t)$, $a_{l,j}(t)$ and $a_{l,k}(t)$.<br>
**solve** equation (2.27).<br>
**if** the solution is unique,<br>
**then** skip to **output**.<br>
**else if** one of the barycentric coordinates $a_{l,x}(t) = 0,\ x \in \{n, h, j, k\}$,<br>
**then** set the sign coefficient $\gamma_{l,x}(t) = 1$ and go back to **solve**.<br>
**else if** two of the barycentric coordinates $a_{l,h}(t) = 0,\ a_{l,k}(t) = 0$<br>
and $d_{jn}(t) + d_{nl}(t) = d_{jl}(t)$,<br>
**then** $\left(\gamma_{l,n}(t), \gamma_{l,h}(t), \gamma_{l,j}(t), \gamma_{l,k}(t)\right) = (1, 1, -1, 1)$.<br>
**else if** two of the barycentric coordinates $a_{l,h}(t) = 0,\ a_{l,k}(t) = 0$<br>
and $d_{jn}(t) + d_{nl}(t) = d_{jl}(t)$,<br>
**then** $\left(\gamma_{l,n}(t), \gamma_{l,h}(t), \gamma_{l,j}(t), \gamma_{l,k}(t)\right) = (1, 1, 1, 1)$.<br>
**else if** two of the barycentric coordinates $a_{l,h}(t) = 0,\ a_{l,k}(t) = 0$<br>
and $d_{jl}(t) + d_{jn}(t) = d_{nl}(t)$,<br>
**then** $\left(\gamma_{l,n}(t), \gamma_{l,h}(t), \gamma_{l,j}(t), \gamma_{l,k}(t)\right) = (-1, 1, 1, 1)$.<br>
**output** $\left(\gamma_{l,n}(t), \gamma_{l,h}(t), \gamma_{l,j}(t), \gamma_{l,k}(t)\right)$.<br>
**end**

---

**Remark 2.3** The main differences of the localization estimation schemes of multi-agent systems with arbitrary deployment in two-dimensional space and three-dimensional space are as follows: (i) The calculation method of the barycentric coordinates. (ii) The determination method of the signed time-varying scalar coefficients. The localization estimation algorithm in three-dimensional space is the same as that in two-dimensional space.

After determining the sign coefficients of the barycentric coordinates, the distributed localization estimation scheme of the multi-agent systems can be designed. The detailed design and analysis process is the same as in Sect. 2.3.2 and is omitted here.

## 2.5 Simulation and Experiment

In this section, the numerical example and the UAV and robot experiment are given to illustrate the effectiveness and the realizability of the proposed methods.

### 2.5.1 Numerical Simulation

(i) Numerical simulation in the 2D plane

In this section, seven agents are taken as simulation objects, and the effectiveness of the proposed localization methods is verified. Two different cases are considered: The first case is that the follower agents are located in the convex hull and the other case is that the follower agents are deployed arbitrarily.

**Case 1** Follower agents are located in the convex hull

The directed graph of the multi-agent systems is shown in Fig. 2.8, where $l_1, l_2, l_3$ are leaders, $f_1, f_2, f_3, f_4$ are followers and the shadow area is a convex hull formed by three leaders. The trajectory curves of seven agents including three leaders and four followers are shown in Fig. 2.9. The initial estimated values of the trajectory curves of all follower agents are given randomly.

Then, applying Algorithm 2.1, the maximum estimation error in the estimation process along the iteration axis is shown by the solid line in Fig. 2.10. The maximum estimation error is defined as $e_{l,\max}(i) = \max_t \left\{ \left\| \mathbf{c}_l^i(t) - \mathbf{c}_i^*(t) \right\|_2 \right\}$. One can conclude that Algorithm 2.1 can guarantee the asymptotic convergence of the estimation error.

In order to show the advantages of the proposed Algorithm 2.1 in convergence rate, a simulation comparison is given. In [13], Ahn et al. proposed a distributed position estimation algorithm based on relative position measurement for the localization estimation problem of single-integrator modeled multi-agent system. The specific algorithm is given below

Fig. 2.8 The directed graph of all agents in Case 1

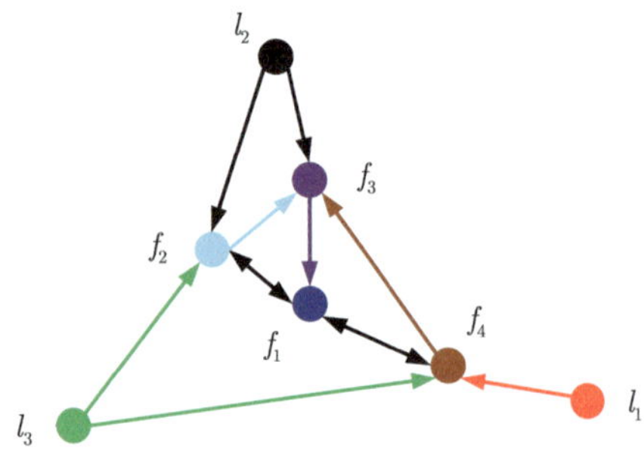

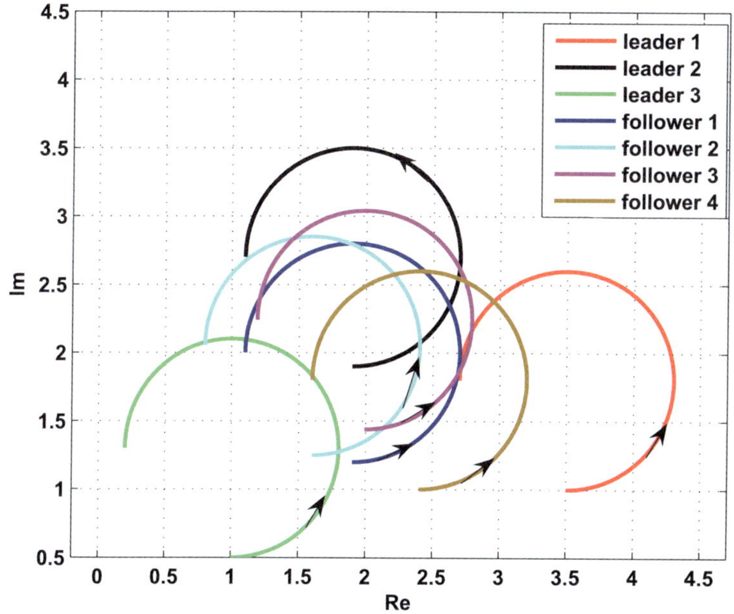


**Fig. 2.9** The trajectories of all agents in Case 1

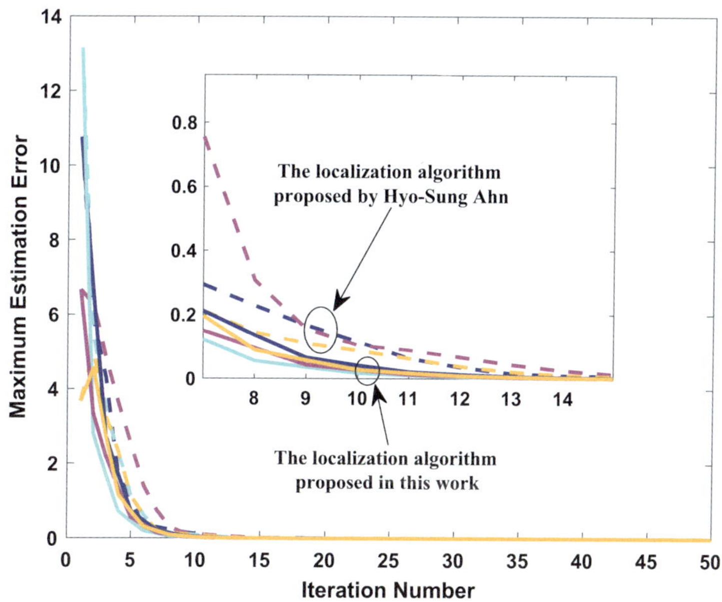


**Fig. 2.10** The comparison of the maximum estimation error under the Algorithm 2.1 and (2.29) in Case 1

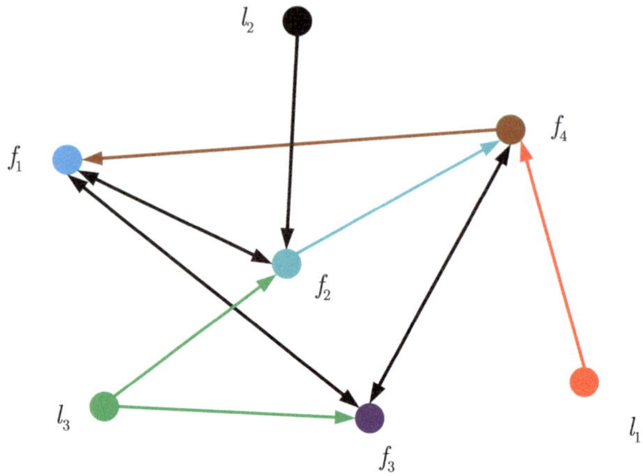


**Fig. 2.11** The directed graph of all agents in Case 2

$$\hat{p}_l^{i+1}(t) = \hat{p}_l^i(t) + k_c u_l^i(t) + k_o \sum_{j \in N_l} a_{lj} \left[ \left( \hat{p}_j^i(t) - \hat{p}_l^i(t) \right) - \left( p_j(t) - p_i(t) \right) \right], \tag{2.29}$$

where $k_c > 0$, $k_o > 0$, $u_l^i(t)$ represents the velocity of the agent $l$, $a_{lj} > 0$ is the weighting factor and $\left(p_j(t) - p_i(t)\right)$ denotes the measured relative position.

The simulation conditions are same with the Algorithm 2.1. Select $k_c = 0.01$, $k_o = 0.4$ and all the weighting factor are set to one. Applying the estimation algorithm (2.29), the simulation results are shown by the dotted line in Fig. 2.10.

Through the simulation results, one can find that the convergence rate of the estimator with Algorithm 2.1 is faster than that under the estimation algorithm (2.29). One can conclude that the proposed distributed localization estimation algorithm based on iterative learning can not only guarantee the accurate localization estimation, but also improve the convergence rate.

**Case 2** Follower agents deployed arbitrarily

The directed graph of the multi-agent systems can be seen in Fig. 2.11, where the shadow area is a convex hull formed by $l_1$, $l_2$ and $l_3$. The trajectory curves of leader agents and follower agents are shown in Fig. 2.12.

The scalar coefficient is selected as $\delta = 0.23$, and applying Algorithm 2.3. Figure 2.13 shows the maximum estimation error curves along the iteration axis. It can be found that the estimation error can converge to zero after 40 iterations, one can draw the conclusion that the proposed Algorithms 2.2 and 2.3 can achieve the real-time localization even the agents are deployed arbitrarily.

Through the simulation results, one can conclude that the proposed two distributed localization estimation algorithms are both effective. One can find that the real-time localization of dynamic multi-agent systems under directed graph is realized.

(ii) Numerical simulation in the 3D space

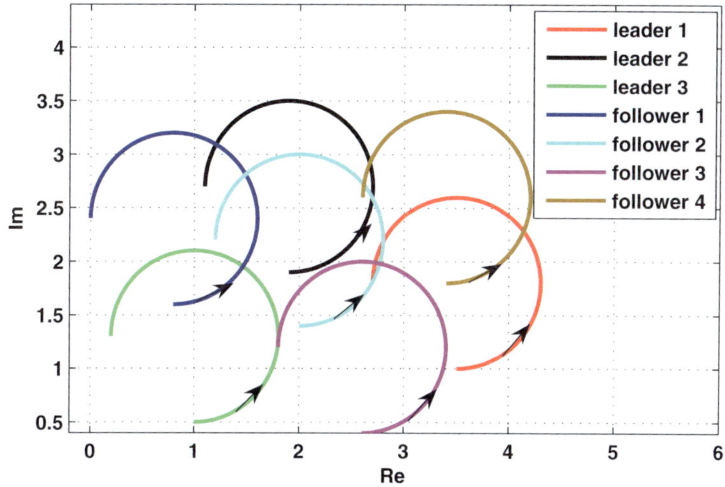


**Fig. 2.12** The trajectories of all agents in Case 2

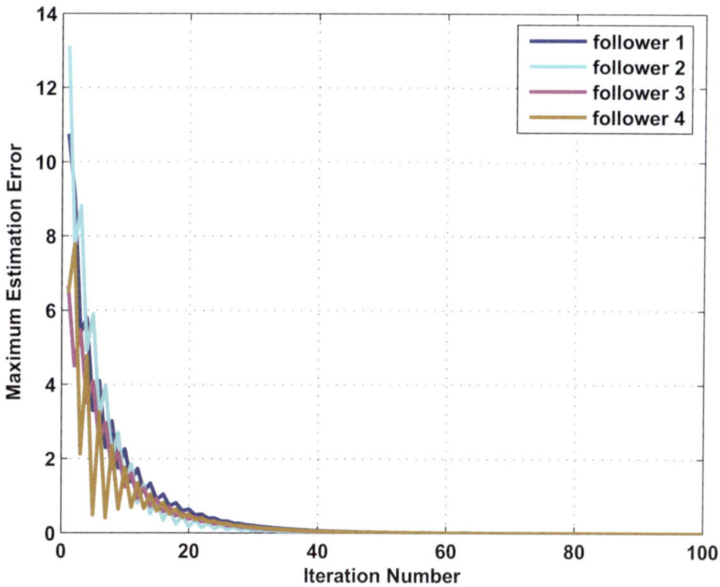


**Fig. 2.13** The maximum estimation error of four follower agents in Case 2

The multi-agent systems with seven agents is taken as the simulation object. The directed communication topology between the seven agents and their initial positions in 3-D space are shown in Fig. 2.14. The real trajectory curve of the leader agent 1 is shown in Fig. 2.15, where the red arrow represents the move direction of the agent.

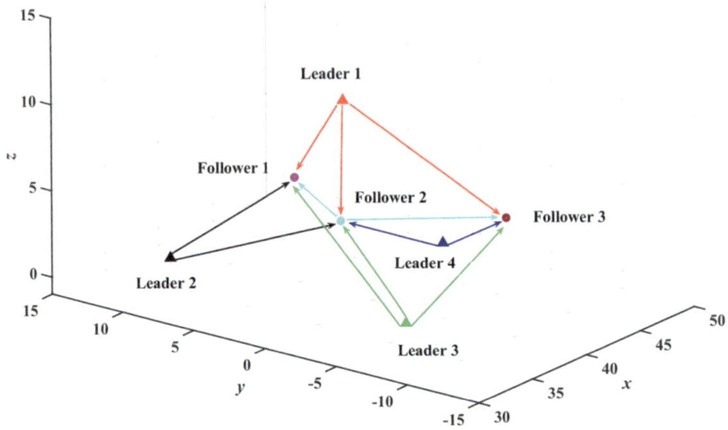


**Fig. 2.14** The directed topology between the seven agents and their initial position in 3-D space

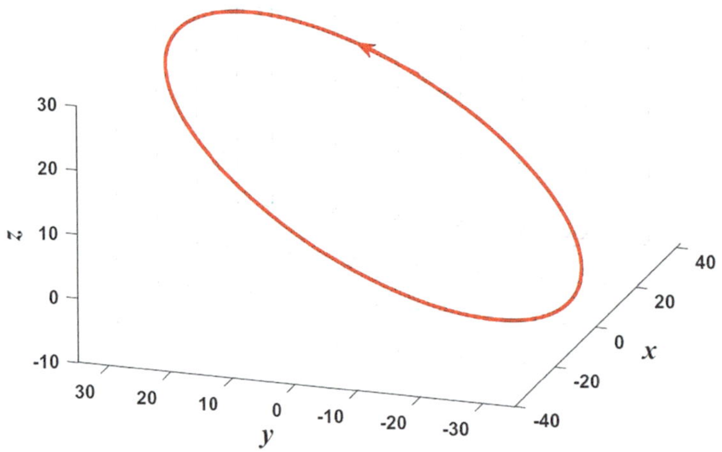


**Fig. 2.15** The trajectory of the agent in 3-D space

The initial estimated values of the follower agents are given randomly. Applying Algorithm 2.4 and (2.17), the maximum estimation errors $e_{l,\max}(i) = \max_t \left\{ \left\| \hat{\mathbf{c}}_l^i(t) - \mathbf{c}_l^*(t) \right\| \right\}$ along the iteration axis are shown in Fig. 2.16. It can be seen that the maximum estimation errors can asymptotically converge to zero.

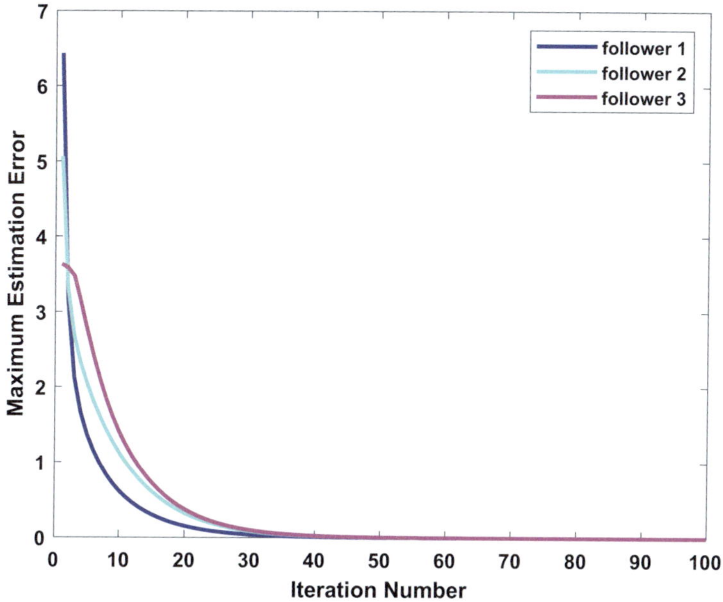


**Fig. 2.16** The maximum estimation error

### *2.5.2 Experiment*

This subsection presents an experimental validation of the proposed localization algorithms in this chapter based on the QBot-2e two-wheeled mobile robot and Quad-rotor UAV.

(i) Qbot-2e robot experiment

In this subsection, the Qbot-2e two-wheeled mobile robot produced by Quanser company are taken as experiment objects to verify the practicality of the proposed methods. A host computer with a Wi-Fi router is used to handle communication between robots. The communication interface between the robot and Matlab is established by the control software QUARC. The experimental platform is shown in Fig. 2.17.

In the experiment, the self-built map and cameras are used to obtain the real coordinates of six robots, but only leader robot can use the real coordinates directly, the follower robot estimates its own coordinates through the proposed methods. The velocity of the robot is 0.7 m/s. The Laser Lidar is used to measure the relative distance between robots and its sampling frequency is 60 Hz.

Two different cases are considered: the follower robots are located in the convex hull and the follower robots are deployed arbitrarily.

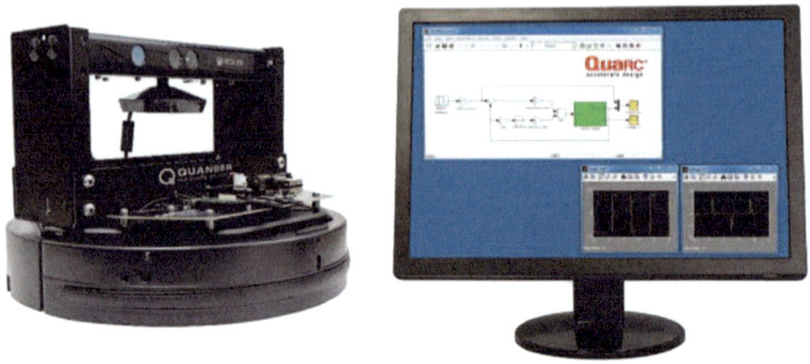


**Fig. 2.17** The QBot-2e robot experimental platform

**Fig. 2.18** The deployment of six robots

**Case 1** Follower robots are located in the convex hull

The deployment of six robots including four leaders and two followers are shown in Fig. 2.18. The four robots in outer circle are leaders, and the other two robots in inner circle are followers. The real trajectory curves of six robots are shown in Fig. 2.19, where one grid is 0.6 m. The initial estimated coordinates of two follower robots are given randomly. Applying Algorithm 2.1, the experimental results are as follows.

The maximum estimation error curves along the iteration axis are shown in Fig. 2.20. One can conclude that the asymptotic convergence of the estimation error can be guaranteed by Algorithm 2.1.

**Case 2** Follower robots deployed arbitrarily

The deployment of six robots including three leaders and three followers are shown in Fig. 2.21. The directed graph of the six robots can be seen in Fig. 2.22,

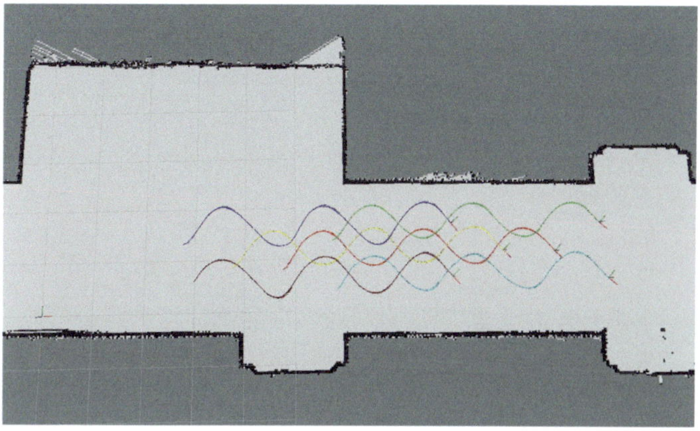

**Fig. 2.19** The real trajectory curves of six robots

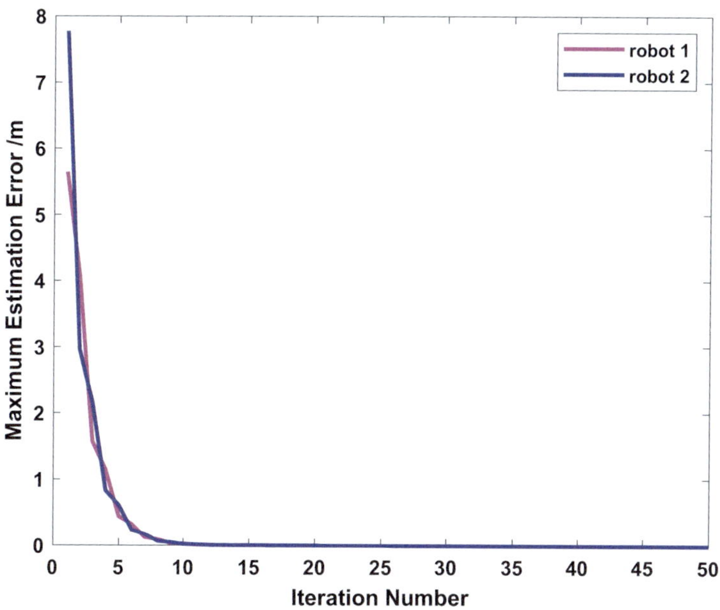


**Fig. 2.20** The maximum estimation error of two follower robots in Case 1

where the shadow area is a convex hull formed by $l_1, l_2, l_3$. The real trajectory curves of the robots are shown in Fig. 2.23.

The initial estimated coordinates of all follower robots are given randomly. The scalar coefficient is selected as $\delta = 0.06$. Applying Algorithm 2.3, the maximum estimation error curves of the two follower robots are illustrated by Fig. 2.24. From the experimental results, one can conclude that the proposed methods not only guarantees

**Fig. 2.21** The deployment of all robots in Case 2

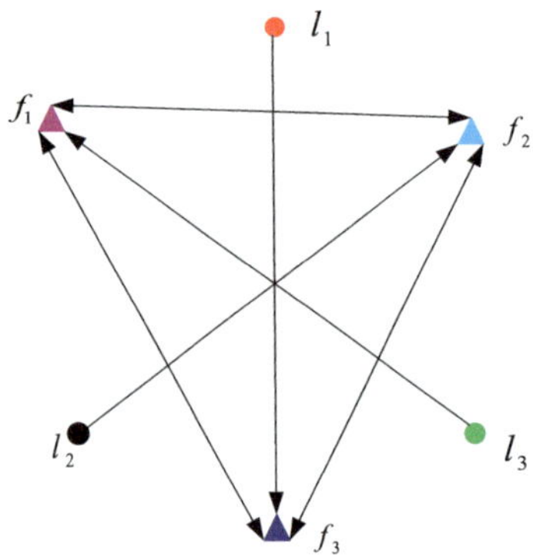


**Fig. 2.22** The directed graph of six robots

the asymptotic convergence of the estimation error, but also realizes the real-time localization of the robots.

(ii) Quad-rotor UAV experiment

In this subsection, we will show the physical experimental platform and conduct the mini quad-rotor UAV experiment to verify the practicability of the Algorithm 2.4 and (2.17). The experimental results will be discussed and analyzed.

OptiTrack system is a high-precision optical motion capture system, which belongs to passive infrared optical positioning technology, as shown in Fig. 2.25. It can capture luminous mark points attached to the quad-rotor UAV. The motion-capture system with 20 cameras is installed in an 8 m × 3 m × 3 m space. The communication between the motion capture system and the base station is based on the protocol defined by Robot Operating System (ROS).

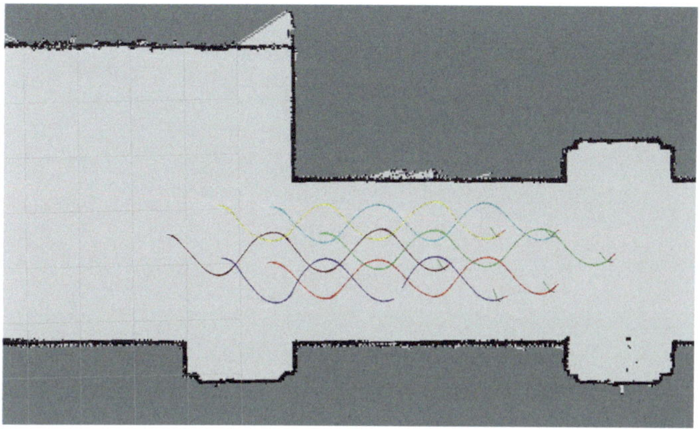

**Fig. 2.23** The real trajectory curves of six robots

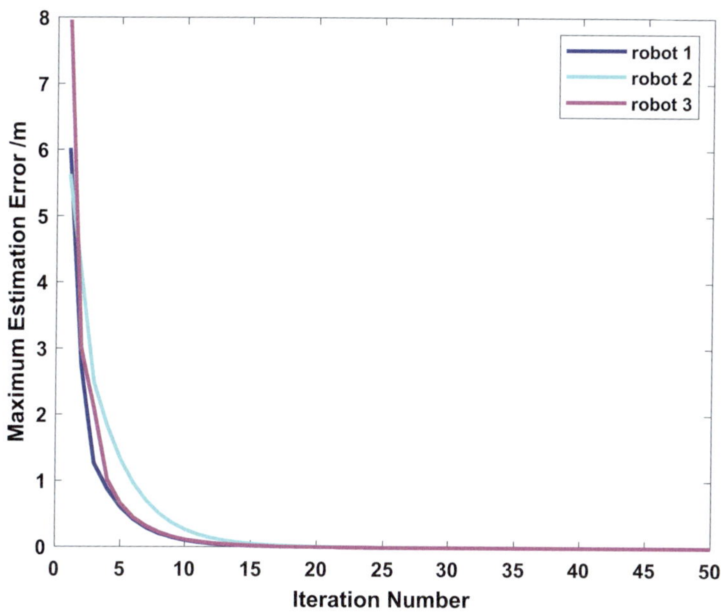


**Fig. 2.24** The maximum estimation error of three follower robots in Case 2

In this experiment, ten mini quad-rotor UAV crazyflies[14] with passive spherical markers are used to test the validity of the Algorithm 2.4 and (2.17). The specific of the mini quad-rotor UAV crazyflie can be seen in Fig. 2.26.

Figure 2.27 shows the deployment of the ten crazyflies. The neighbor sets of the six follower UAVs are given as follows: $N_{F1} = \{F3, F5, L2, L3\}$, $N_{F2} = \{F4, L1, L2, L4\}$, $N_{F3} = \{L1, L2, L3, L4\}$, $N_{F4} = \{F1, L2, L3, L4\}$,

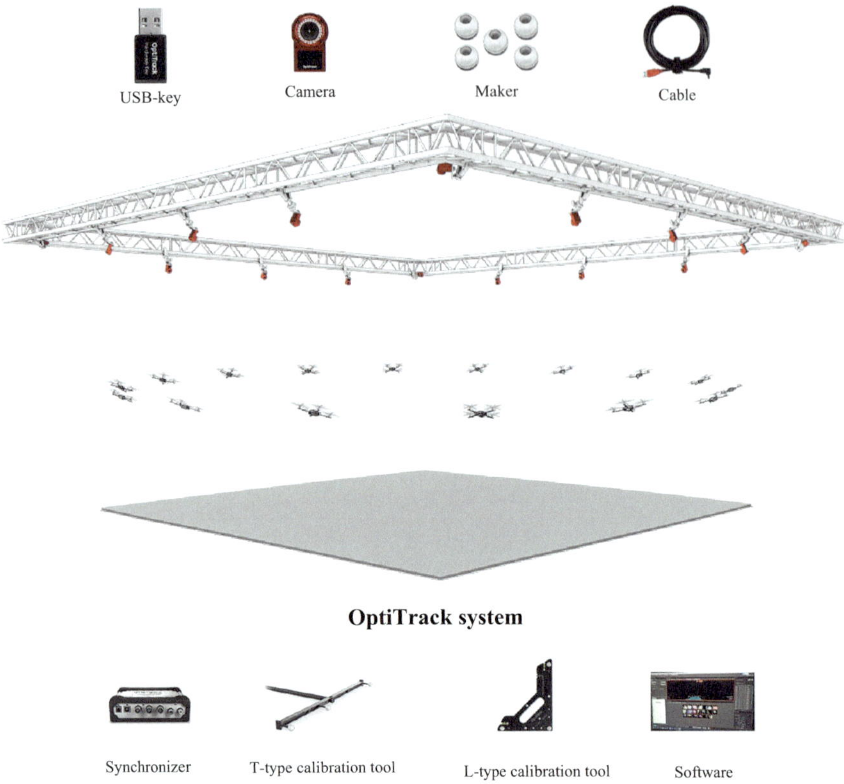


**Fig. 2.25** The high-precision optical motion capture system

$N_{F5} = \{F2, F3, L2, L4\}$, $N_{F6} = \{F2, L1, L3, L4\}$, where $L$ represents the leader and $F$ denotes the follower. The real-time positions of the ten UAVs are shown in Fig. 2.28. The real coordinates of the crazyflie is obtained by the OptiTrack system, the leaders know their real coordinates, and the followers estimate their coordinates through the estimation method. In the experiment, Ultra-Wideband (UWB) technology is used to measure the relative distance between the UAVs.

The initial estimated values of the follower UAVs are given randomly. Select $\delta = 3.2$, applying Algorithm 2.4 and (2.17), the maximum estimation errors are shown in Fig. 2.29, where the unit is meter. It can be found that the accurate localization can be guaranteed after applying the proposed Algorithm 2.4 and (2.17).

From the simulation and experimental results, the following conclusion can be drawn: The real-time localization problem of multi-agent systems with arbitrary deployment in 2D plane and 3D space is solved under Algorithms 2.1–2.4.

**Fig. 2.26** The mini quad-rotor UAV crazyflies

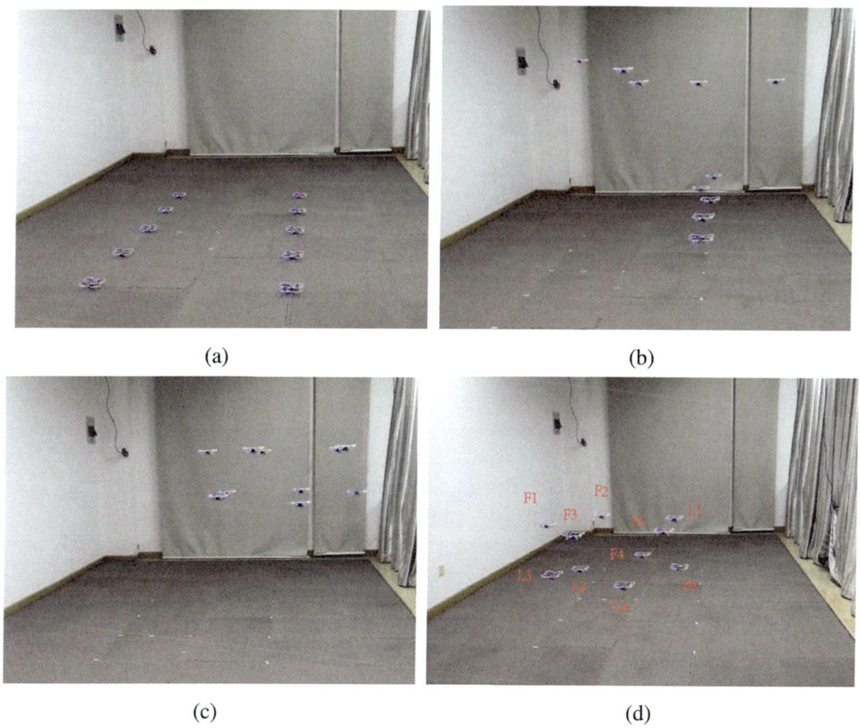


**Fig. 2.27** The formation flight process of the ten crazyflies

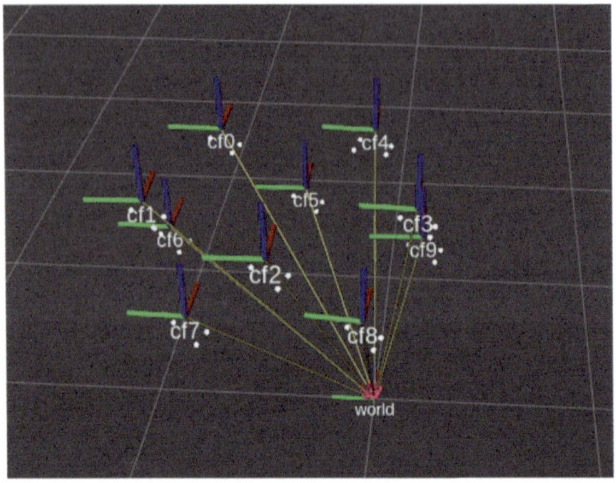


**Fig. 2.28** The real-time position of the crazyflies in experiment

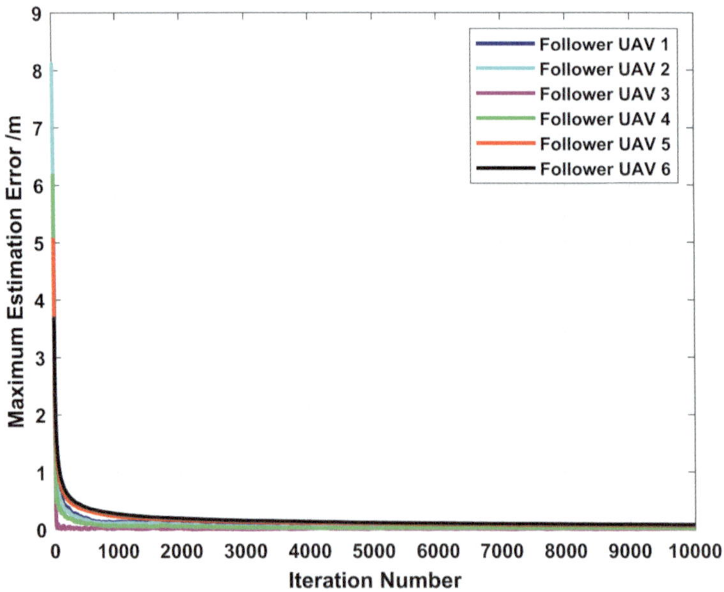


**Fig. 2.29** The maximum estimation error of the ten crazyflies

## 2.6 Conclusions

This chapter investigates the problem of accurate localization in multi-agent systems with repetitive operational characteristics in two-dimensional planes and three-dimensional spaces. For the localization problem of agents within a convex hull, a distributed localization estimation algorithm based on iterative learning is proposed, where the estimated coordinates at the same sampling time in the previous iteration are used to update the current iteration's coordinates, enabling precise localization estimation of the agents. The results are further extended to multi-agent systems with arbitrary deployment, and a new computation and determination mechanism is designed to establish the sign coefficients. An improved distributed localization estimation algorithm with sign coefficients is proposed, achieving precise localization of agents. By utilizing the infinity norm and the Richardson iteration method, global asymptotic convergence of the multi-agent system under random initial conditions is guaranteed. The effectiveness and practicality of the algorithm are validated through simulations and experiments.

## References


1. Khan, U., Kar, S., Moura, J.: Distributed sensor localization in random environments using minimal number of anchor nodes. IEEE Trans. Signal Process **57**(5), 2000–2016 (2009)
2. Diao, Y., Lin, Z., Fu, M.: A barycentric coordinate based distributed localization algorithm for sensor networks. IEEE Trans. Signal Process **62**(18), 4760–4771 (2014)
3. Wang, Z., Chang, Y., Zhang, H., et al.: Bearing-only distributed localization for multi-agent systems with complex coordinates. Inform. Sci. **626**, 837–850 (2023)
4. Chen, L.: Triangular angle rigidity for distributed localization in 2D. Automatica **143**, 110414 (2022)
5. Deghat, M., Shames, I., Anderson, B., et al.: Localization and circumnavigation of a slowly moving target using bearing measurements. IEEE Trans. Autom. Control **59**(8), 2182–2188 (2014)
6. Deghat, M., Xia, L., Anderson, B., et al.: Multi-target localization and circumnavigation by a single agent using bearing measurements. Int. J. Robust Nonlinear Control **25**(14), 2362–2374 (2015)
7. Wang, Z., Shen, J., Fu, X., et al.: Robust adaptive optimization algorithm for stability control of electric patrol survey UAV, Proc. Int. Conf. Comput. Intell. Inf. Syst., 108–115 (2017)
8. Yufka, A., Parlaktuna, O., Ozkan, M.: Formation-based cooperative transportation by a group of non-holonomic mobile robots, IEEE Int. Conf. Syst. Man Cybern., **12**(4), 3300–3307 (2010)
9. Ahn, H. S., Moore, K. L., Chen, Y. : Trajectory-keeping in satellite formation flying via robust periodic learning control, Int. J. Robust Nonlinear Control, **20**(14), 1655–1666 (2010)
10. Arimoto, S., Kawamura, S., Miyazaki, F.: Bettering operation of robots by learning. J. Robot. Syst. **1**(2), 123–140 (1984)
11. Richardson, L.F.: The approximate arithmetical solution by finite differences of physical problems involving differential equations, with an application to the stresses in a masonry dam. Philos. Trans. Roy. Soc. Lond. Series A, Containing Papers Math. Phys. Char. **210**, 307–357 (1911)
12. Varga, R.: Matrix Iterative Analysis. Springer (1962)
13. Oh, K., Ahn, H.: Formation control of mobile agents based on distributed position estimation. IEEE Tran. Autom. Control **58**(3), 737–742 (2013)
14. Bandala, A., Nakano, R., Faelden, G., et al.: Implementation of an artificial neural network in recognizing in-flight quadrotor images. In: Proceedings of TENCON 2015-2015 IEEE Region 10 Conference, Macao, China, pp. 1–5 (2015)

# Chapter 3
# Distributed Localization of MASs with Imperfect Channels

The real-time localization problem for a class of multi-agent systems with imperfect channels under directed graphs is investigated in this chapter. A successive distributed localization estimation algorithm based on iterative learning is proposed to solve the localization problem of the agent with bounded communication delays. The barycentric coordinates are used to represent the relative position between agents. An estimator is proposed to inspect whether the received data packet is the latest through the iteration stamp on the packet. Meanwhile, an intermittent distributed localization algorithm is devised to handle with the random data loss of agents in real-time localization problem. The relationship between the convergence rate and the probability of the data loss is given in the sense of mathematical expectations. The numerical simulation is provided to verify the effectiveness of the aforementioned methods.

## 3.1 Introduction

In the past three decades, there are rich literature on accurate localization algorithm for wireless sensor networks. In [1], based on the measurement of relative distance, the node localization problem is transformed into the solution of pseudo linear system. The problem of bearing-based network localization is addressed [2]. Besides, three-dimensional network localization problem with mixed local relative measurements is studied [3]. It should be mentioned that, the aforementhioned results focused on the localization over static networks, however the real-time localization over dynamically changing networks is more common in real applications. Some efforts have been made on the distributed localization for dynamically changing networks. In [4], the localization and circumnavigation problem for a group of agents

 
Y. Lv et al., *Distributed Localization and Formation Control of Multi-Agent Systems with Local Sensing Information*, https://doi.org/10.1007/978-981-96-6015-5_3

using bearing measurements are investigated. In three dimensional space, the localization and circumnavigation problems of the multi-agent systems for an unknown stationary target is solved [5].

In some complex network environments, the data communication delay between agents is inevitable. The reason for data communication delay usually lies in the imperfect channels of the wireless networks, such as traffic congestion, instability and bandwidth constraints, and so on. Until recent years, the distributed localization problem with data communication delay begins to receive attention. In [6], the convergence of the localization algorithm can be guaranteed under the measurement delay only if the GPS information is added together with the relative measurements. The global convergence of asynchronous localization algorithm under communication delay and packet loss is proved via the extended delay graph theory [7, 8]. Although the consensus of multi-agent systems with communication delay has received a lot of attention, the problem of real-time localization of multi-agent systems with communication delay has been largely overlooked.

To be specific, based on iterative learning, we proposed a successive distributed localization estimation algorithm to surmount the localization problem for the agent with communication delay. Meanwhile, the localization estimation of agent with random data loss is also taken into consideration. In addition, in terms of the stochastic technology, the relationship between the convergence rate and the probability of the data loss is analysed.

Different from the localization estimation of static sensor networks, the real-time localization estimation of multi-agent systems is realized in this chapter. Compared with the existing distributed localization algorithms for dynamically changing networks with perfect channels [9, 10], assumptions that agents running in a given way and moving slowly are removed. Besides, the real-time localization problem of the multi-agent systems without velocity constraint is also addressed. Unlike the intermittent control scheme proposed for the networked systems with imperfect channels [11, 12], the proposed successive algorithm can guarantee that the position estimation process is not interrupted, which can improve the convergence speed of the estimator. The main contribution of this work are summarized as follows.

(1) The real-time localization of multi-agent systems with communication delay and data loss under directed graphs is realized.
(2) The proposed distributed algorithm based on iterative learning solves the localization problem of multi-agent systems without velocity constraint.
(3) The proposed successive localization algorithm improves the convergence speed of the estimation error.

The rest of this chapter is arranged as follows: Sect. 3.2 shows the problem descriptions. Section 3.3 gives the estimator design and convergence analysis when the network is communication delays. Section 3.4 extends the result to the multi-agent systems with random data loss. In Sect. 3.5, some simulation results are shown to verify the effectiveness of the proposed algorithm. Finally, Sect. 3.6 concludes this work.

## 3.2 Problem Formulation

A $N$-node multi-agent systems that can perform task repeatedly is considered, where $m$ agents are the leaders and other $(N - m)$ agents are the followers. The leaders can obtain their own position coordinates through location devices. The followers can only estimate their position by exchanging information with their neighbors.

In the process of position estimation, the agent needs to communicate with each other to transfer their own estimated coordinates. If the communication channels is perfect, the packets can be delivered on time and synchronously. However, communication delay and random data dropout is inevitable when encountered with imperfect channels. An example can be seen in Fig. 3.1, where $i$ represents the iteration number. In Fig. 3.1a, Agent 1 transmits its estimated value to Agent 2, where the red solid circle indicates the packet to be transmitted and the solid line with arrow indicates the direction of transmission. From Fig. 3.1a, one can find that the data packet transmitted later may arrive earlier than the data packet transmitted earlier, and the data packet transmission delay will continue for a period of time. In Fig. 3.1b, the red hollow circle indicates that the data packet is lost during transmission.

The objective is to develop a distributed localization scheme for the multi-agent systems with communication delay and random data loss. The developed localization algorithm can guarantee the estimation error $\|\mathbf{e}_l(t)\|_\infty$ converges to zero, where $\mathbf{e}_l(t)=\hat{\mathbf{c}}_l(t) - \mathbf{c}_l^*(t)$, $\hat{\mathbf{c}}_l(t)$ is the estimated coordinate and $\mathbf{c}_l^*(t)$ represents the real coordinate.

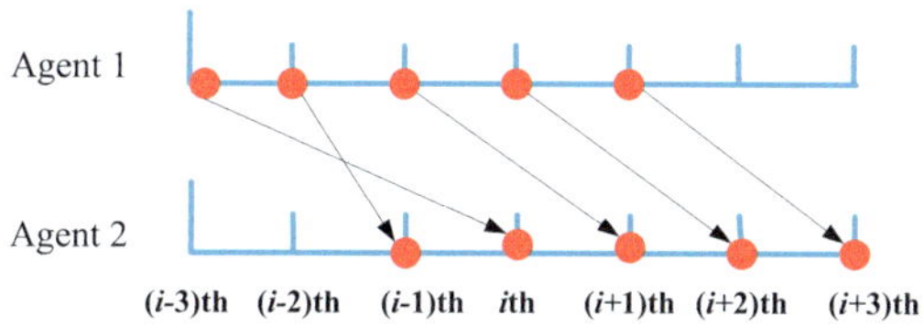


(a) The communication delay between agents

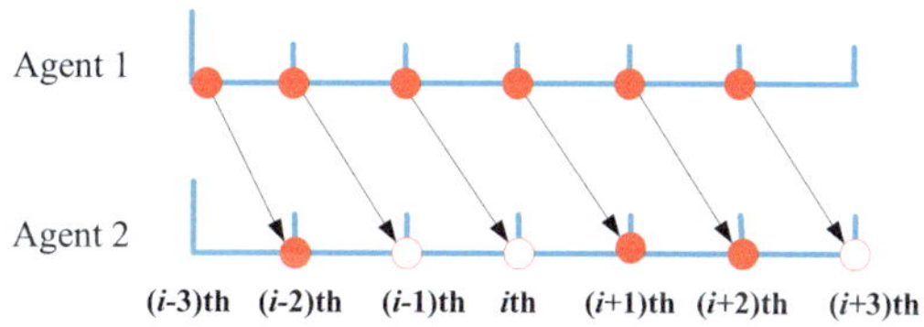


(b) The random data dropout between agents

**Fig. 3.1** The communication delay and the random data dropout between agents

## 3.3 Distributed Localization Algorithm with Bounded Communication Delays

The arrival time of packets with different iteration stamps will be chaotic when the communication delay occurs, as shown in Fig. 3.1a. Consider the communication delay between the follower agents, the estimated coordinates received by the follower agent may not be the latest. The objective of this section is designing a successive distributed localization algorithm to solve the real-time localization problem of the multi-agent systems with bounded communication delays.

### *3.3.1 Estimator Design*

If there is no communication delay and data loss between the agents, for the follower agent $l$, a distributed localization algorithm based on iterative learning is designed as follows,

$$\hat{\mathbf{c}}_l^{i+1}(t) = \sum_{q \in \mathcal{N}_l} a_{l,q}^{i+1}(t)\hat{\mathbf{c}}_q^i(t) + \sum_{w \in \mathcal{N}_l} a_{l,w}^{i+1}(t)\mathbf{c}_w(t), \tag{3.1}$$

where $i$ represents the repeated operation number, here we call it iteration number. $t = 0, 1, \ldots, T$ is the sampling time with $T$ being the terminal time instant. $\mathcal{N}_l$ is the set of neighbors of the agent $l$, $\mathbf{c}_w(t)$ and $\hat{\mathbf{c}}_q^i(t)$ denote the coordinates of the leader neighbor and the follower neighbor, respectively. Note that, since a leader agent can obtain its coordinates directly, then there is no need to add iterative index $i$ to $\mathbf{c}_w(t)$ and moreover, $c_w(t) = c_w^*(t)$, where $c_w^*(t)$ represents the real coordinates.

**Remark 3.1** In (3.1), the estimated coordinates $\hat{\mathbf{c}}_q^i(t)$ of the $i$th iteration and the real coordinates $\mathbf{c}_w(t)$ are used to update the estimated coordinates $\hat{\mathbf{c}}_l^{i+1}(t)$ of the $(i+1)$th iteration. Moreover, only the local information from its neighboring agents is available when the follower agent estimates its coordinates, so the localization algorithm (3.1) is distributed.

For all $(N - m)$ followers, one has a compact form

$$\hat{\mathbf{C}}_s^{i+1}(t) = \mathbf{P}^{i+1}(t)\hat{\mathbf{C}}_s^i(t) + \mathbf{B}^{i+1}(t)\mathbf{C}_a(t), \tag{3.2}$$

where $\hat{\mathbf{C}}_s^i(t) = \left[\left(\hat{\mathbf{c}}_{m+1}^i(t)\right)^{\mathrm{T}}, \ldots, \left(\hat{\mathbf{c}}_N^i(t)\right)^{\mathrm{T}}\right]^{\mathrm{T}}$ are the estimated coordinates of all the followers, and $\mathbf{C}_a(t) = \left[(\mathbf{c}_1(t))^{\mathrm{T}}, \ldots, (\mathbf{c}_m(t))^{\mathrm{T}}\right]^{\mathrm{T}}$ denotes the real coordinates of all the leaders.

$$\mathbf{P}^{i+1}(t)=\begin{bmatrix} 0 & a^{i+1}_{m+1,m+2}(t) & \cdots & a^{i+1}_{m+1,N}(t) \\ a^{i+1}_{m+2,m+1}(t) & 0 & \cdots & a^{i+1}_{m+2,N}(t) \\ \vdots & \vdots & \ddots & \vdots \\ a^{i+1}_{N,m+1}(t) & a^{i+1}_{N,m+2}(t) & \cdots & 0 \end{bmatrix} \in R^{(N-m)\times(N-m)}$$ characterizes the transmission of information between the followers.

$$\mathbf{B}^{i+1}(t)=\begin{bmatrix} a^{i+1}_{m+1,1}(t) & a^{i+1}_{m+1,2}(t) & \cdots & a^{i+1}_{m+1,m}(t) \\ a^{i+1}_{m+2,1}(t) & a^{i+1}_{m+2,2}(t) & \cdots & a^{i+1}_{m+2,m}(t) \\ \vdots & \vdots & \ddots & \vdots \\ a^{i+1}_{N,1}(t) & a^{i+1}_{N,2}(t) & \cdots & a^{i+1}_{N,m}(t) \end{bmatrix} \in R^{(N-m)\times m}$$ characterizes the transmission of information between the followers and the leaders.

In order to determine whether the current received data is the latest, each data packet has an iteration stamp. One can guarantee that the stored data is latest by inspecting the iteration stamp. A successive distributed localization estimation scheme is shown in Algorithm 3.1.

---

**Algorithm 3.1**. Estimate the real coordinate of the follower agent under bounded communication delays

---

**for** $i = 1 : Iteration.$

**initialization** Given the initial iteration estimated coordinates $\hat{\mathbf{c}}^i_l(t)$, $l \in \{m+1, \ldots, N\}$ of the follower agent. Set $l_i = 1$, where $l_i$ is a data tag in memory.

**Step 1.** Each agent transmits its own estimated coordinates packet $\Phi\left(i, \hat{\mathbf{c}}^i_l(t)\right)$ to its neighbors, where $\Phi\left(i, \hat{\mathbf{c}}^i_l(t)\right)$ is the estimated coordinates packet.

**Step 2.** if $i > l_i$ or $l_i = 1$, then store $\Phi\left(i, \hat{\mathbf{c}}^i_l(t)\right)$ and update $l_i := i$

else discard packet $\Phi\left(i, \hat{\mathbf{c}}^i_l(t)\right)$

end

**Step 3.** Measure the relative distance, and calculate the barycentric coordinate.

**Step 4.** Update

$$\hat{\mathbf{c}}^{i+1}_l(t) = \sum_{q\in N_l} a^{l_i+1}_{l,q}(t)\hat{\mathbf{c}}^{l_i}_q(t) + \sum_{w\in N_l} a^{i+1}_{l,w}(t)\mathbf{c}_w(t)$$

**Step 5.** Update $i = i + 1$ and return to **Step 1**.

**end**

---

**Remark 3.2** Algorithm 3.1 implies that when agent $l$ updates $\hat{\mathbf{c}}^{i+1}_l(t)$ in the $(i+1)$th iteration, it does not only use the $i$th iteration estimated coordinates of the neighboring agents, but also uses the previous iteration estimated coordinates which are stored in the memory. In other words, even if agent $l$ cannot receive the $i$th iteration estimated coordinates transmitted by the neighbors, the position estimation process is not interrupted. Therefore, Algorithm 3.1 is called a successive estimation scheme.

**Remark 3.3** The iteration stamp is used to judge whether the received data is latest or not. Therefore, asymptotic convergence of the estimation error is guaranteed. If there is no iteration stamp on the packet, i.e., as long as the data packet is received, the estimator is updated, no matter whether the data is the latest or not. The estimation error will fluctuate greatly in the estimation process, and the convergence speed will slow down. Detailed verification and analysis are given in the simulation.

Next, the mathematical model of Algorithm 3.1 is given below. Variable $\tau_{lj}^{i}(t)$ denote the communication delay when agent $l$ transmits its estimated position coordinates to agent $j$. The following assumption will be used.

**Assumption 3.1** For any $(v_l, v_j) \in \varepsilon$, there exits a constant $\tau_d < \infty$ satisfies $\tau_{lj}^{i}(t) < \tau_d$.

Denote by $\tau_{\max}$ the maximum communication delay between agents. Based on Algorithm 3.1, a mathematical model of the successive distributed localization algorithm with bounded communication delays is proposed.

$$\begin{aligned} \hat{\mathbf{C}}_s^{i+1}(t) =& \bar{\mathbf{P}}_0^{i+1}(t)\hat{\mathbf{C}}_s^{i}(t) + \bar{\mathbf{P}}_1^{i}(t)\hat{\mathbf{C}}_s^{i-1}(t) + \cdots + \\ & \bar{\mathbf{P}}_{\tau_{\max}}^{i+1-\tau_{\max}}(t)\hat{\mathbf{C}}_s^{i-\tau_{\max}}(t) + \mathbf{B}^{i+1}(t)\mathbf{C}_a(t), \end{aligned} \tag{3.3}$$

where matrixes $\bar{\mathbf{P}}_0^{i+1}(t) \in R^{(N-m)\times(N-m)}, \ldots, \bar{\mathbf{P}}_{\tau_{\max}}^{i+1-\tau_{\max}}(t) \in R^{(N-m)\times(N-m)}$ represent the relative position between the follower agents at different iterations. Due to the existence of communication delays, the estimated coordinates received by the follower agent in the $(i+1)$th iteration include not only the $i$th iteration result of its neighbors, but also the previous iteration results.

### 3.3.2 Convergence Analysis

The convergence of the localization algorithm in the presence of communication delay is provided.

**Theorem 3.1** *Suppose that the multi-agent systems perform tasks repeatedly. Algorithm 3.1 solves the real-time localization problem with bounded communication delays, i.e.,* $\lim_{i\to\infty} \left\|\mathbf{E}_s^{i+1}(t)\right\|_\infty = 0$.

***Proof*** For the convenience of analysis, estimation algorithm (3.3) is reformulated as follows

$$\hat{\mathbf{C}}_s^{i+1}(t) = \left[\bar{\mathbf{P}}_0^{i+1}(t), \ldots, \bar{\mathbf{P}}_{\tau_{\max}}^{i+1-\tau_{\max}}(t)\right] \begin{bmatrix} \hat{\mathbf{C}}_s^{i}(t) \\ \vdots \\ \hat{\mathbf{C}}_s^{i-\tau_{\max}}(t) \end{bmatrix} + \mathbf{B}^{i+1}(t)\mathbf{C}_a(t). \tag{3.4}$$

Subtracting $\mathbf{C}_s^*(t)$ from both sides of (3.4), one has

$$\mathbf{E}_s^{i+1}(t) = \left[\bar{\mathbf{P}}_0^{i+1}(t), \ldots, \bar{\mathbf{P}}_{\tau_{\max}}^{i+1-\tau_{\max}}(t)\right] \begin{bmatrix} \mathbf{E}_s^i(t) \\ \vdots \\ \mathbf{E}_s^{i-\tau_{\max}}(t) \end{bmatrix} + \left(\bar{\mathbf{P}}_0^{i+1}(t) + \cdots + \bar{\mathbf{P}}_{\tau_{\max}}^{i+1-\tau_{\max}}(t) - \mathbf{I}\right) \mathbf{C}_s^*(t) + \mathbf{B}^{i+1}(t)\mathbf{C}_a^*(t). \tag{3.5}$$

Since the trajectory of the agent is repeated along the iteration axis, the sum of the corresponding barycentric coordinates with different iteration superscripts in the matrixes $\bar{\mathbf{P}}_0^{i+1}(t), \ldots, \bar{\mathbf{P}}_{\tau_{\max}}^{i+1-\tau_{\max}}(t)$ are equal to the corresponding barycentric coordinates in matrix $\mathbf{P}^{i+1}(t)$. Then one can conclude that $\bar{\mathbf{P}}_0^{i+1}(t) + \cdots + \bar{\mathbf{P}}_{\tau_{\max}}^{i+1-\tau_{\max}}(t) = \mathbf{P}^{i+1}(t)$.

According to the above analysis, $\left(\bar{\mathbf{P}}_0^{i+1}(t) + \cdots + \bar{\mathbf{P}}_{\tau_{\max}}^{i+1-\tau_{\max}}(t) - \mathbf{I}\right) \mathbf{C}_s^*(t) + \mathbf{B}^{i+1}(t)\mathbf{C}_a^*(t) = \mathbf{0}$ holds.

Taking infinite norms on both sides of (3.5), one has

$$\left\|\mathbf{E}_s^{i+1}(t)\right\|_\infty \le \left\|\left[\bar{\mathbf{P}}_0^{i+1}(t), \ldots, \bar{\mathbf{P}}_{\tau_{\max}}^{i+1-\tau_{\max}}(t)\right]\right\|_\infty \left\|\begin{bmatrix} \mathbf{E}_s^i(t) \\ \vdots \\ \mathbf{E}_s^{i-\tau_{\max}}(t) \end{bmatrix}\right\|_\infty. \tag{3.6}$$

Since $\bar{\mathbf{P}}_0^{i+1}(t) + \cdots + \bar{\mathbf{P}}_{\tau_{\max}}^{i+1-\tau_{\max}}(t) = \mathbf{P}^{i+1}(t)$ and the element in matrix $\mathbf{P}^{i+1}(t)$ represents the barycentric coordinates between the follower agents, the sum of each row of matrix $\mathbf{P}^{i+1}(t)$ is less than 1. Then, according to the property of infinite norm, $\left\|\left[\bar{\mathbf{P}}_0^{i+1}(t), \ldots, \bar{\mathbf{P}}_{\tau_{\max}}^{i+1-\tau_{\max}}(t)\right]\right\|_\infty = \sigma_2 < 1$.

Based on the analysis from (3.5) to (3.6), one can obtain that

$$\left\|\mathbf{E}_s^{i+1}(t)\right\|_\infty \le \sigma_2 \left\|\begin{bmatrix} \mathbf{E}_s^i(t) \\ \vdots \\ \mathbf{E}_s^{i-\tau_{\max}}(t) \end{bmatrix}\right\|_\infty. \tag{3.7}$$

By solving (3.7), one has

$$\left\|\mathbf{E}_s^{i+1}(t)\right\|_\infty \le \sigma_2^{i+1-\tau_{\max}} \left\|\begin{bmatrix} \mathbf{E}_s^{\tau_{\max}}(t) \\ \vdots \\ \mathbf{E}_s^0(t) \end{bmatrix}\right\|_\infty. \tag{3.8}$$

Since $0 < \sigma_2 < 1$ and the estimation error $\left\|\begin{bmatrix} \mathbf{E}_s^{\tau_{\max}}(t) \\ \vdots \\ \mathbf{E}_s^0(t) \end{bmatrix}\right\|_\infty$ is bounded, then (3.8) implies that $\lim_{i\to\infty} \left\|\mathbf{E}_s^{i+1}(t)\right\|_\infty = 0$.

The proof of Theorem 3.1 is completed.

## 3.4 Distributed Localization Algorithm with Random Data Loss

In this section, the random data loss between agents is considered, as shown in Fig. 3.1b. The objective of this section is designing a distributed localization algorithm to solve the localization problem with random data loss.

### 3.4.1 Estimator Design

Consider the random data loss between agents as shown in Fig. 3.1b. An intermittent distributed localization estimation algorithm based on iterative learning is proposed as follows

$$\begin{aligned}&\hat{\mathbf{c}}_l^{i+1}(t) = (1-\alpha_l^{i+1}(t))\hat{\mathbf{c}}_l^i(t)+\\&\alpha_l^{i+1}(t)\left(\sum_{q\in\mathcal{N}_l} a_{l,q}^{i+1}(t)\hat{\mathbf{c}}_q^i(t) + \sum_{w\in\mathcal{N}_l} a_{l,w}^{i+1}(t)\mathbf{c}_w(t)\right),\end{aligned} \tag{3.9}$$

where the stochastic parameter $\alpha_l^{i+1}(t)$ represents the independent and identically distributed Bernoulli random variable. Its value is either 1 or 0, which characterizes whether the agent $l$ receives data transmitted by its' neighbors.

The probability of $\alpha_l^{i+1}(t)$ is define as : prob $\left\{\alpha_l^{i+1}(t)=1\right\} = \mathbb{E}\left\{\alpha_l^{i+1}(t)\right\} = \bar{\alpha}_l$, $0 < \bar{\alpha}_l < 1$.

The process of intermittent distributed localization algorithm is presented in Algorithm 3.2.

**Algorithm 3.2**. Estimate the real coordinate of the follower agent under random data loss

**for** $i = 1 : Iteration.$

**initialization** Given the initial iteration estimated coordinates $\hat{\mathbf{c}}_l^i(t)$, $l \in \{m+1, \ldots, N\}$ of the follower agent.

**Step 1.** Each agent transmits its own estimated coordinates $\hat{\mathbf{c}}_l^i(t)$ to its neighbors.

**Step 2.** Measure the relative distance, and calculate the barycentric coordinate.

**Step 3.** if the transmission is successful, then update $\hat{\mathbf{c}}_l^{i+1}(t) = \sum_{q\in\mathcal{N}_l} a_{l,q}^{i+1}(t)\hat{\mathbf{c}}_q^i(t) + \sum_{w\in\mathcal{N}_l} a_{l,w}^{i+1}(t)\mathbf{c}_w(t)$

end.

else if, the transmission is failure, then stop updating and keep the current estimated coordinates $\hat{\mathbf{c}}_l^i(t)$ unchanged.

end.

**Step 4.** Update $i = i+1$ and return to **Step 1**.

**end**

**Remark 3.4** In (3.9), if the data transmission is successful, the coordinates $\hat{\mathbf{c}}_l^{i+1}(t)$ of agent $l$ is updated by the estimated coordinates of the $i$th iteration transmitted by its neighbors, and the probability is $\bar{\alpha}_l$. If the data transmission is failure, then agent $l$ keeps the estimated value $\hat{\mathbf{c}}_l^i(t)$ unchanged, and the probability is $1-\bar{\alpha}_l$.

**Remark 3.5** After introducing Bernoulli random variables, Algorithm 3.2 can also be used to describe the localization of the multi-agent systems with bounded communication delays. In other words, if there is no delay in data transmission, normal update is performed. And if the data cannot be received on time, the update is suspended. Thus, Algorithm 3.2 is general.

### 3.4.2 Convergence Analysis

The convergence of the localization algorithm in the presence of data loss is provided.

**Theorem 3.2** *Suppose that the multi-agent systems perform tasks repeatedly. Algorithm 3.2 solves the real-time localization problem with random data loss, i.e., if the probability of packet loss $\bar{\alpha}_l > 0$ is satisfied, one can guarantee that* $\lim_{i\to\infty} \mathbb{E}\left\{\left\|\mathbf{E}_s^{i+1}(t)\right\|_\infty\right\} = 0$.

***Proof*** For all $(N-m)$ followers, one has

$$\hat{\mathbf{C}}_s^{i+1}(t) = \left(\mathbf{I}-\boldsymbol{\alpha}^{i+1}(t)\right)\hat{\mathbf{C}}_s^i(t) + \boldsymbol{\alpha}^{i+1}(t)\left(\mathbf{P}^{i+1}(t)\hat{\mathbf{C}}_s^i(t) + \mathbf{B}^{i+1}(t)\mathbf{C}_a(t)\right), \tag{3.10}$$

where $\boldsymbol{\alpha}^{i+1}(t) = \begin{bmatrix} \alpha_{m+1}^{i+1}(t) & 0 & \cdots & 0 \\ 0 & \alpha_{m+2}^{i+1}(t) & \cdots & 0 \\ \vdots & \vdots & \ddots & \vdots \\ 0 & 0 & \cdots & \alpha_N^{i+1}(t) \end{bmatrix}$.

Subtracting $\mathbf{C}_s^*(t)$ from both sides of (3.10), and according to the definition of estimation error, one can obtain that

$$\begin{aligned} \mathbf{E}_s^{i+1}(t) &= \left[\left(\mathbf{I}-\boldsymbol{\alpha}^{i+1}(t)\right) + \boldsymbol{\alpha}^{i+1}(t)\mathbf{P}^{i+1}(t)\right]\mathbf{E}_s^i(t) \\ &+\boldsymbol{\alpha}^{i+1}(t)\left(\left(\mathbf{P}^{i+1}(t)-\mathbf{I}\right)\mathbf{C}_s^*(t) + \mathbf{B}^{i+1}(t)\mathbf{C}_a^*(t)\right) \\ &+\boldsymbol{\alpha}^{i+1}(t)\mathbf{B}^{i+1}(t)\mathbf{E}_a(t). \end{aligned} \tag{3.11}$$

Taking infinite norm on both sides of (3.11), one has

$$\begin{aligned} \left\|\mathbf{E}_s^{i+1}(t)\right\|_\infty &\le \left\|\left(\mathbf{I}-\boldsymbol{\alpha}^{i+1}(t)\right) + \boldsymbol{\alpha}^{i+1}(t)\mathbf{P}^{i+1}(t)\right\|_\infty \\ &\times\left\|\mathbf{E}_s^i(t)\right\|_\infty + \left\|\boldsymbol{\alpha}^{i+1}(t)\mathbf{B}^{i+1}(t)\right\|_\infty \|\mathbf{E}_a(t)\|_\infty \\ &+\left\|\boldsymbol{\alpha}^{i+1}(t)\left(\left(\mathbf{P}^{i+1}(t)-\mathbf{I}\right)\mathbf{C}_s^*(t) + \mathbf{B}^{i+1}(t)\mathbf{C}_a^*(t)\right)\right\|_\infty. \end{aligned} \tag{3.12}$$

It suffices to show that $\left\|\boldsymbol{\alpha}^{i+1}(t)\left(\left(\mathbf{P}^{i+1}(t)-\mathbf{I}\right)\mathbf{c}_s^*(t)+\mathbf{B}^{i+1}(t)\mathbf{c}_a^*(t)\right)\right\|_\infty$ is equal to zero from above analysis. Since the leader agent can obtain its coordinate value directly, $\|\mathbf{E}_a(t)\|_\infty$ is equal to zero no matter the data transmission is successful or failure.

Taking mathematical expectation on both sides of (3.12), one has

$$\mathbb{E}\left\{\left\|\mathbf{E}_s^{i+1}(t)\right\|_\infty\right\} \leq \left\|(\mathbf{I}-\bar{\boldsymbol{\alpha}})+\bar{\boldsymbol{\alpha}}\mathbf{P}^{i+1}(t)\right\|_\infty \mathbb{E}\left\{\left\|\mathbf{E}_s^{i}(t)\right\|_\infty\right\}. \tag{3.13}$$

The detail of $(\mathbf{I}-\bar{\boldsymbol{\alpha}})+\bar{\boldsymbol{\alpha}}\mathbf{P}^{i+1}(t)$ is shown as below

$$(\mathbf{I}-\bar{\boldsymbol{\alpha}})+\bar{\boldsymbol{\alpha}}\mathbf{P}^{i+1}(t)=\begin{bmatrix} 1-\bar{\alpha}_{m+1} & \begin{matrix}\bar{\alpha}_{m+1}\times\\ a^{i+1}_{m+1,m+2}(t)\end{matrix} & \cdots & \begin{matrix}\bar{\alpha}_{m+1}\times\\ a^{i+1}_{m+1,N}(t)\end{matrix} \\ \begin{matrix}\bar{\alpha}_{m+2}\times\\ a^{i+1}_{m+2,m+1}(t)\end{matrix} & 1-\bar{\alpha}_{m+2} & \cdots & \begin{matrix}\bar{\alpha}_{m+2}\times\\ a^{i+1}_{m+2,N}(t)\end{matrix} \\ \vdots & \vdots & \ddots & \vdots \\ \begin{matrix}\bar{\alpha}_{N}\times\\ a^{i+1}_{N,m+1}(t)\end{matrix} & \begin{matrix}\bar{\alpha}_{N}\times\\ a^{i+1}_{N,m+2}(t)\end{matrix} & \cdots & 1-\bar{\alpha}_N \end{bmatrix}. \tag{3.14}$$

Since $0<\bar{\alpha}_l<1$, one can guarantee that all elements in $(\mathbf{I}-\bar{\boldsymbol{\alpha}})+\bar{\boldsymbol{\alpha}}\mathbf{P}^{i+1}(t)$ are positive. According to the definition of infinite norm, $\left\|(\mathbf{I}-\bar{\boldsymbol{\alpha}})+\bar{\boldsymbol{\alpha}}\mathbf{P}^{i+1}(t)\right\|_\infty$ can be expressed as $1-\bar{\alpha}_l+\bar{\alpha}_l\sum_{q\in\mathcal{N}_l} a^{i+1}_{l,q}(t)$, $l\in\{m+1,\dots N\}$. $a^{i+1}_{l,q}(t)$ represents the barycentric coordinate between the follower agents, thus $\sum_{q\in\mathcal{N}_l} a^{i+1}_{l,q}(t)<1$. Based on the above analysis, one can obtain that $1-\bar{\alpha}_l+\bar{\alpha}_l\sum_{q\in\mathcal{N}_l} a^{i+1}_{l,q}(t)<1$. Therefore, $\left\|(\mathbf{I}-\bar{\boldsymbol{\alpha}})+\bar{\boldsymbol{\alpha}}\mathbf{P}^{i+1}(t)\right\|_\infty=\sigma_3<1$.

Based on the analysis from (3.10) to (3.14), one can obtain that

$$\mathbb{E}\left\{\left\|\mathbf{E}_s^{i+1}(t)\right\|_\infty\right\} \leq \sigma_3^{i+1}\mathbb{E}\left\{\left\|\mathbf{E}_s^{0}(t)\right\|_\infty\right\}. \tag{3.15}$$

Since $0<\sigma_3<1$ and the estimation error $\left\|\mathbf{E}_s^0(t)\right\|_\infty$ is bounded, (3.15) implies $\lim_{i\to\infty}\mathbb{E}\left\{\left\|\mathbf{E}_s^{i+1}(t)\right\|_\infty\right\}=0$.

The proof of Theorem 3.2 is completed.

### *3.4.3 Discussion of the Convergence Rate*

In this subsection, the relation between the data loss probability and the convergence rate of the estimator is discussed.

According to the property of infinite norm, one has $\left\|(\mathbf{I}-\bar{\boldsymbol{\alpha}})+\bar{\boldsymbol{\alpha}}\mathbf{P}^{i+1}(t)\right\|_{\infty}=\delta_l=1-\bar{\alpha}_l+\bar{\alpha}_l\sum\limits_{q\in N_l}a_{l,q}^{i+1}(t)$, $l\in\{m+1,\ldots N\}$, and its value will affect the convergence speed of the estimation error. Since $\sum\limits_{q\in N_l}a_{l,q}^{i+1}(t)<1$, when $\bar{\alpha}_l$ decreases, i.e., the probability of packet loss increases, one can find that $\delta_l$ also increases, the value of $\delta_l$ will approach 1. Therefore, the convergence speed of the estimation error will slow down. In contrast, when $\bar{\alpha}_l$ increases, i.e., the probability of packet loss decreases, $\delta_l$ will decrease. The convergence speed of the estimation error will become faster.

From the above analysis, one can draw the conclusion that the data packet loss will not affect the convergence of the estimator, but it will affect the convergence speed. The performance of the estimator will become poor when the probability of the data loss increases.

## 3.5 Numerical Simulation

In this section, the numerical simulation is provided to testify the effectiveness and realizability of the proposed methods.

A multi-agent system with seven agents performing tasks repeatedly is considered. The communication topology between the agents is shown in Fig. 3.2, where the red dots are the leader agents, and the blue dots are the follower agents. The trajectory curves of all the agents are shown in Fig. 3.3, where the follower agents are located in

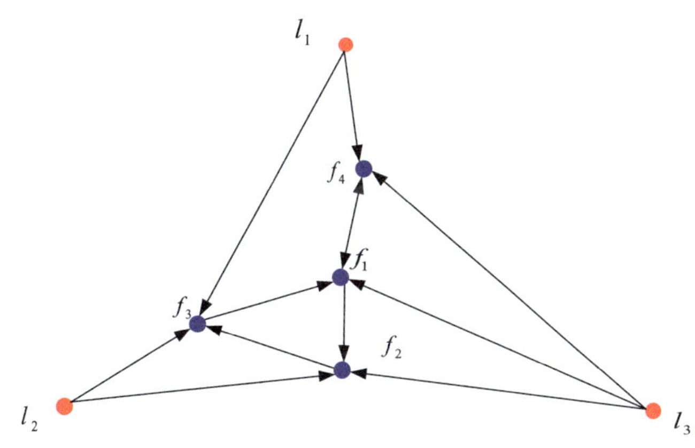


**Fig. 3.2** The directed topology of the seven agents

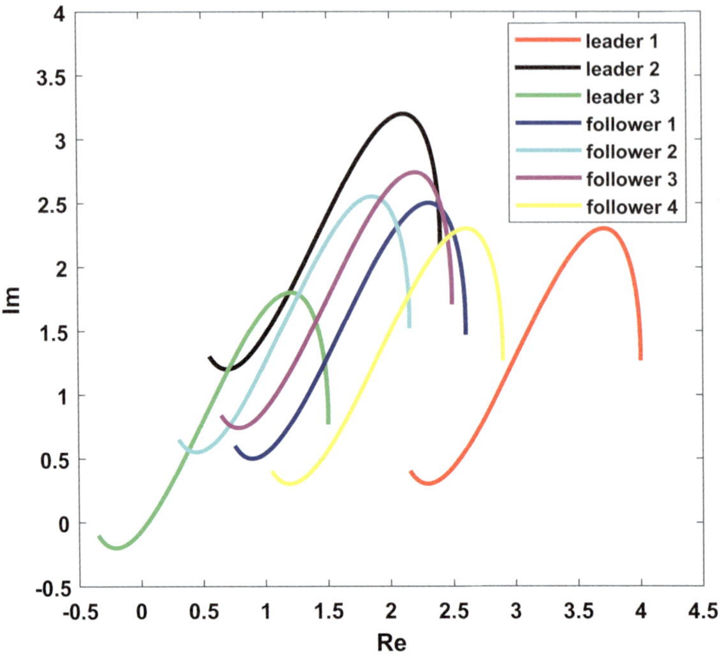


**Fig. 3.3** The real trajectories of the seven agents

the convex hull formed by the leaders. This subsection will discuss the convergence of the Algorithms 3.1 and 3.2 in two cases. The first case is the bounded communication delays. The second case is the data packet loss.

**Case 1** Bounded communication delays.

The communication delay is random and bounded by $\tau_{\max} = 6$. The initial estimated coordinates of the followers are given randomly. Applying Algorithm 3.1, the results are shown in Figs. 3.4 and 3.5.

Figure 3.4 shows the maximum estimation error $e_{l,\max}(i) = \max_t \left\{ \left\| \hat{\mathbf{c}}_l^i(t) - \mathbf{c}_l^*(t) \right\|_2 \right\}$ along the iteration axis. One can find that the estimation error asymptotically converges to zero. The maximum estimation error curves under ideal network transmission are shown in Fig. 3.5.

Figure 3.6 are the maximum estimation error curves without discriminating iteration stamps. It can be found that there are some fluctuations in the convergence process and the convergence speed is slower than that of the Algorithm 3.1. From Figs. 3.4, 3.5 and 3.6, one can conclude that although the convergence speed in the case of bounded communication delay is slower than that in perfect channels, Algorithm 3.1 can guarantee the smooth and gradual convergence of the estimation error without large fluctuation.

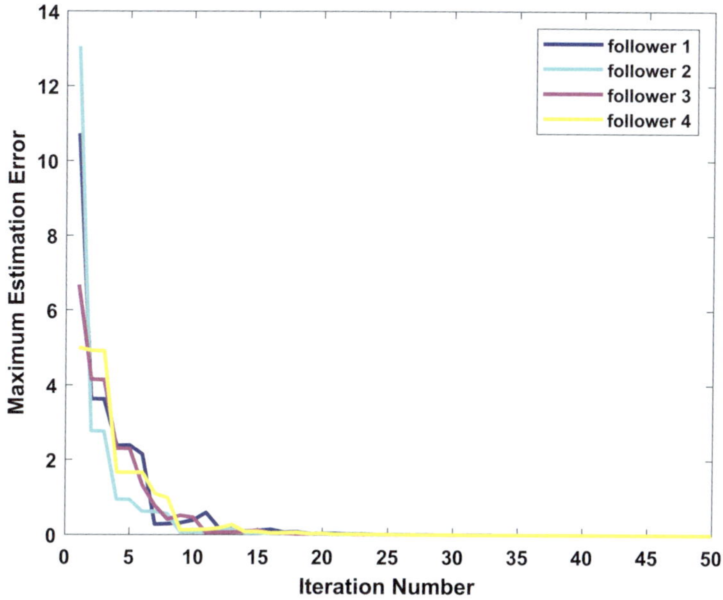


**Fig. 3.4** The maximum estimation error under bounded communication delay

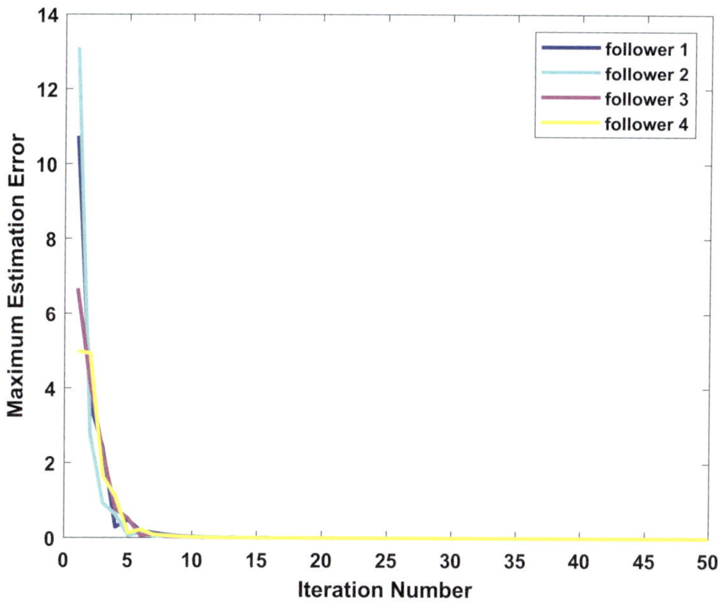


**Fig. 3.5** The maximum estimation error under ideal network transmission

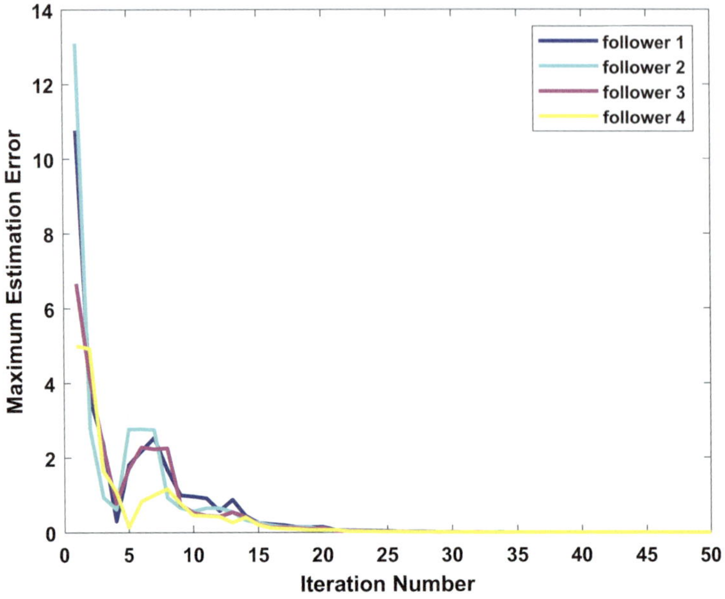


**Fig. 3.6** The maximum estimation error without discriminating iteration stamps

**Case 2** Data packet loss.

In this case, the 10% packet loss probability is considered. The "binopdf" function in MATLAB is used to describe the packet loss probability. The initial estimated coordinates of the followers are given randomly. Applying Algorithm 3.2, the results are shown in Fig. 3.7.

Figure 3.7 shows the maximum estimation error curves under the 10% packet loss probability. It can be found that Algorithm 3.2 guarantees the asymptotic convergence of the estimation error, but the convergence speed is slower than that in perfect channels. Although the initial trajectory curve is far from the actual trajectory, after about 30 iterations, the accurate localization estimation is realized.

Through the simulation study, one can draw the conclusion that the proposed estimation methods can realize the accurate real-time localization of the multi-agent systems with imperfect channels.

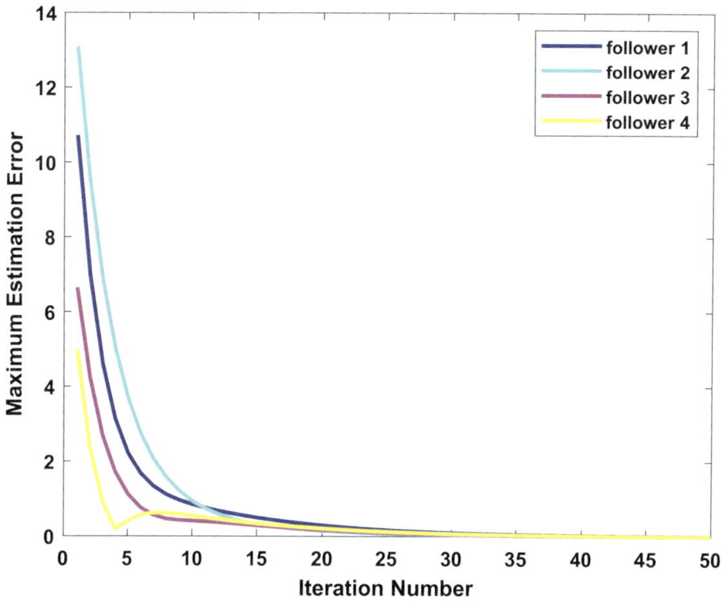


**Fig. 3.7** The maximum estimation error under 10% packet loss probability

## 3.6 Conclusions

The real-time localization problem of the multi-agent systems with imperfect channels is studied in this chapter. The proposed successive distributed localization scheme based on iterative learning solves the localization problem of agents with bounded communication delays. Meanwhile, an intermittent distributed localization algorithm is designed to handle with the localization problem under random data loss. The relationship between the convergence rate and the probability of the data loss is proved. The simulation study verifies the effectiveness of the proposed schemes. The performance of the localization algorithm in random noisy environments is our future research interest.

## References


1. Khan, U., Kar, S., Moura, J.: Distributed sensor localization in random environments using minimal number of anchor nodes. IEEE Trans. Signal Process. **57**(5), 2000–2016 (2009)
2. Zhao, S., Zelazo, D.: Localizability and distributed protocols for bearing-based network localization in arbitrary dimensions. Automatica **69**, 334–341 (2016)
3. Fang, X., Li, X., Xie, L.: 3-D distributed localization with mixed local relative measurements. IEEE Trans. Signal Process. **68**(6), 5869–5881 (2020)

4. Zou, Y., Wang, L., Meng, Z.: Distributed localization and circumnavigation algorithms for a multiagent system with persistent and intermittent bearing measurements. IEEE Trans. Control Syst. Technol. **29**(5), 2092–2101 (2021)
5. Li, R., Shi, Y., Song, Y.: Localization and circumnavigation of multiple agents along an unknown target based on bearing-only measurement: a three-dimensional solution. Automatica **94**, 18–25 (2018)
6. Amato, E., Notaro, I., Mattei, M., et al.: Attitude and position estimation for an UAV swarm using consensus Kalman filtering. IEEE Metrol. Aerosp. 519–524 (2015)
7. Huang, X., Tian, Y.: Asynchronous distributed localization in networks with communication delays and packet losses. Automatica **96**, 134–140 (2018)
8. Huang, X., Tian, Y.: Localization in sensor networks with communication delays and package losses. In: Proceedings of IEEE Conference on Decision and Control, pp. 3974–3979, Melbourne, Australia (2017)
9. Deghat, M., Shames, I., Anderson, B., et al.: Localization and circumnavigation of a slowly moving target using bearing measurements. IEEE Trans. Autom. Control **59**(8), 2182–2188 (2014)
10. Deghat, M., Xia, L., Anderson, B., et al.: Multi-target localization and circumnavigation by a single agent using bearing measurements. Int. J. Robust Nonlinear Control **25**(14), 2362–2374 (2015)
11. Liu, C., Xu, J., Wu, J.: Iterative learning control for remote control systems with communication delay and data dropout. Math. Prob. Eng. 1–14 (2012)
12. Zhang, Y., Liu, J., Ruan, X.: Networked iterative learning control for systems with packet dropout and communication delay. IFAC-Pap. Online **52**(24), 117–122 (2019)

# Chapter 4
# Distributed Localization of MASs with Random Noise

This chapter is concerned with the real-time localization problem for the dynamic multi-agent systems with measurement and communication noises under directed graphs. The barycentric coordinates are introduced to describe the relative position between agents. A novel robust distributed localization estimation algorithm based on iterative learning is proposed. The relative-distance unbiased estimator constructed from the historical iterative information is used to suppress the measurement noise. The designed stochastic approximation method with two iterative-varying gains is used to inhibit the communication noise. Under the zero-mean and independent distributed conditions on the measurement and communication noises, the asymptotic convergence of the proposed methods is derived. The numerical simulation and the QBot-2e robot experiment are conducted to test and verify the effectiveness and the practicability of the proposed methods.

## 4.1 Introduction

In most practical applications, especially in networked environments, measurement noise and communication noise are ubiquitous. The reason for such noise usually results from the occurrence of some random events in the process of measurement or network transmission, such as low sensor accuracy, random failure and recovery of transmission link, sudden environment changes, and so on. The environment is assumed to be ideal in the aforementioned distributed localization algorithms. Thus, it is significance to eliminate the influence of noise on localization accuracy and realize the accurate localization in the presence of noise. Not until very recently, does the distributed localization estimation problem under the measurement noise and communication noise begin to be reported. The modified barycentric-coordinate based localization methods are proposed in [1], which can achieve the accurate



Y. Lv et al., *Distributed Localization and Formation Control of Multi-Agent Systems with Local Sensing Information*, https://doi.org/10.1007/978-981-96-6015-5_4

node-localization in the presence of distance measurement noise. Further, the relative position-based node-localization problem for static network in the presence of measurement and communication noises is investigated in [2]. These localization methods considering noise are aimed at static sensor networks, and cannot be directly applied to the real-time localization problem of the dynamic multi-agent systems investigated in this paper. However, the real-time localization of the multi-agent systems with random noise is very important for the implementation of its follow-up tasks and the existing related works are few. Therefore, it is meaningful and challenging to design a novel robust distributed estimation algorithm to realize the accurate localization of the agent, which is the third motivation of this work.

To be specific, a robust distributed localization estimation algorithm based on iterative learning is proposed, which can solve the real-time accurate localization problem of multi-agent systems with measurement and communication noises. The main innovations of this work are threefold:

(1) The real-time localization problem of the multi-agent systems with repetitive characteristics under the directed graph is solved.
(2) A novel estimation method based on iterative learning is proposed, thus the velocity constraint is removed.
(3) The proposed robust distributed localization algorithm solves the real-time accurate localization problem of the multi-agent systems with both the measurement noise and the communication noise.

The rest of this chapter is arranged as follows: Sect. 4.2 shows the preliminary and the problem formulation. The estimator design and convergence analysis in the presence of measurement and communication noises are presented in Sect. 4.3. A corollary is proposed in Sect. 4.4. In Sect. 4.5, some simulation and experimental results are provided. Finally, Sect. 4.6 concludes this work.

## 4.2 Problem Formulation

A N-node dynamic multi-agent system with $m$ leaders and $(N-m)$ followers is considered, which can perform task repeatedly. The leaders obtain their coordinates directly through the location device. The followers estimate their coordinates by exchanging information with the leaders.

The measured relative distance is used to calculate the barycentric coordinates. Affected by the complexity of the application environment and the sensor measurement accuracy, the measured relative distance between agents contains noise. Meanwhile, in the process of position estimation, the agent needs to exchange its estimated coordinates with neighbors through the wireless network. In practice, communication channels are inevitably interfered by stochastic noises. The measurement noise and the communication noise caused by various uncertainties in the data measuring and transmission processes may degrade the localization accuracy.

The objective of this chapter is to develop an efficient robust distributed localization estimation method for the dynamic multi-agent systems with measurement noise and communication noise. The developed estimation algorithm can achieve accurate estimation, i.e., $\|\mathbf{e}_l(t)\|_\infty = 0$, where $\mathbf{e}_l(t) = \hat{\mathbf{c}}_l(t) - \mathbf{c}^*(t)$ is the estimation error, $\hat{\mathbf{c}}_l(t)$ represents the estimated coordinate and $\mathbf{c}^*(t)$ denotes the real coordinate.

## 4.3 Distributed Localization Algorithm with Measurement Noise and Communication Noise

Consider the case that the follower agent is located in the convex hull formed by three of its neighbors as shown in Fig. 1.6a. The localization problem of the multi-agent systems with measurement noise and communication noise is considered. The objective of this section is to design a novel robust distributed localization estimation algorithm based on iterative learning to solve the real-time accurate localization problem.

The measured relative distance at the $t$th sampling time of the $i$th iteration is

$$\tilde{d}^i_{xy}(t) = d^i_{xy}(t) + \upsilon^i_{xy}(t), \tag{4.1}$$

where $i$ is the iteration number, $d^i_{xy}(t)$ denotes the real distance between the agent $x$ and the agent $y$, $\upsilon^i_{xy}(t)$ represents the distance measurement noise.

An assumption for the measurement noise is given below.

**Assumption 4.1** For all $x,\ y \in \upsilon$, $i \in \mathbb{N}$, the distance measurement noise $\upsilon^i_{xy}(t)$ is bounded and independent identically distributed along the iterative axis, that is the mathematical expectation of $\upsilon^i_{xy}(t)$ is equal to zero.

Before designing the estimation algorithm, the estimated relative distance $\hat{d}^i_{xy}(t)$ is obtained according to the following method.

$$\begin{aligned} \hat{d}^i_{xy}(t) &= \tfrac{1}{i}\sum_{n=0}^{i} \tilde{d}^n_{xy}(t) = \tfrac{i-1}{i}\hat{d}^{i-1}_{xy}(t) + \tfrac{1}{i}\tilde{d}^i_{xy}(t), \\ \hat{d}^0_{xy}(t) &= \tilde{d}^0_{xy}(t). \end{aligned} \tag{4.2}$$

According to Assumption 4.1 and the strong law of large numbers [3]. One can obtain that the estimated relative distance $\hat{d}^i_{xy}(t)$ converges to $d^i_{xy}(t)$ when iteration number $i$ goes to infinity, i.e., $\lim\limits_{i\to\infty} \hat{d}^i_{xy}(t) = d^i_{xy}(t)$.

### 4.3.1 Estimator Design

A robust distributed localization estimation algorithm based on iterative learning is proposed as follows

$$\hat{\mathbf{c}}_l^{i+1}(t) = (1-\alpha^{i+1}(t))\hat{\mathbf{c}}_l^i(t) + \beta^{i+1}(t)\left(\sum_{q\in N_l}\hat{a}_{l,q}^{i+1}(t)(\hat{\mathbf{c}}_q^i(t)+\varphi_q^i(t)) + \sum_{w\in N_l}\hat{a}_{l,w}^{i+1}(t)(\mathbf{c}_w^*(t)+\varphi_w^i(t))\right), \tag{4.3}$$

where $N_l$ denotes the set of neighbors of the agent $l$; $\mathbf{c}_w^*(t)$ represents the real coordinates of the leader neighbor agent and $\hat{\mathbf{c}}_q^i(t)$ represents the estimated coordinates of the follower neighbor agent. $\alpha^{i+1}(t) = \frac{1}{(1+i)^{\lambda_1}}$ and $\beta^{i+1}(t) = \frac{1}{(1+i)^{\lambda_2}}$ are two iteration-varying gains, $0 < \lambda_1 < \lambda_2 < 1$, $\lambda_2 - \lambda_1 < \zeta$ and $\zeta > 0$ is a small bounded value. $\hat{a}_{l,q}^{i+1}(t)$ and $\hat{a}_{l,w}^{i+1}(t)$ are the estimated barycentric coordinates. $\varphi_q^i(t)$ and $\varphi_w^i(t)$ represent the communication noises in different communication channel.

The leader agent can obtain its own coordinates directly through the location device, so there is no iterative index $i$ on $\mathbf{c}_w^*(t)$.

**Assumption 4.2** For all $q, w \in N_l$, $i \in \mathbb{N}$, $\varphi_q^i(t)$ and $\varphi_w^i(t)$ are the variables of bounded and independent identically distributed along the iterative axis.

The process of distributed localization estimation algorithm based on iterative learning is presented in Algorithm 4.1.

---

**Algorithm 4.1**. Estimate the real coordinate of the follower agent with measurement and communication noises

---

**for** $i = 1 : Iteration.$

**initialization** Given the initial iteration estimated coordinates $\hat{\mathbf{c}}_l^i(t)$, $l \in \{m+1, \ldots, N\}$ and the constant gains $\lambda_1$ and $\lambda_2$.

**Step 1.** Each agent sends its own estimated coordinates to its neighbors.

**Step 2.** Measure the relative distance $\tilde{d}_{xy}^i(t)$ and calculate the estimated relative distance $\hat{d}_{xy}^i(t)$:

$$\hat{d}_{xy}^i(t) = \frac{1}{i}\sum_{n=0}^{i}\tilde{d}_{xy}^n(t) = \frac{i-1}{i}\hat{d}_{xy}^{i-1}(t) + \frac{1}{i}\tilde{d}_{xy}^i(t).$$

**Step 3.** Calculate the barycentric coordinates $\hat{a}_{l,q}^{i+1}(t)$ and $\hat{a}_{l,w}^{i+1}(t)$ according to Cayley-Menger determinant.

**Step 4.** Update

$$\hat{\mathbf{c}}_l^{i+1}(t) = (1-\alpha^{i+1}(t))\hat{\mathbf{c}}_l^i(t) + \beta^{i+1}(t)\left(\sum_{q\in N_l}\hat{a}_{l,q}^{i+1}(t)(\hat{\mathbf{c}}_q^i(t)+\varphi_q^i(t)) + \sum_{w\in N_l}\hat{a}_{l,w}^{i+1}(t)(\mathbf{c}_w^*(t)+\varphi_w^i(t))\right).$$

**Step 5.** Update $i = i + 1$ and return to **Step 1**.

**end**

---

**Remark 4.1** The barycentric coordinates $\hat{a}_{l,q}^{i+1}(t)$ and $\hat{a}_{l,w}^{i+1}(t)$ are calculated based on the estimated relative distance $\hat{d}_{xy}^{i+1}(t)$. There are some deviation between the estimated barycentric coordinates $\hat{a}_{l,q}^{i+1}(t)$, $\hat{a}_{l,w}^{i+1}(t)$ and the real barycentric coordinates $a_{l,q}^{i+1}(t)$, $a_{l,w}^{i+1}(t)$ in the presence of measurement noise. According to the strong law of large numbers, the estimated barycentric coordinates is equal to the real barycentric coordinates when $i$ goes to infinity, i.e., $\lim_{i\to\infty} \hat{a}_{l,q}^{i+1}(t) = a_{l,q}^{i+1}(t)$ and $\lim_{i\to\infty} \hat{a}_{l,w}^{i+1}(t) = a_{l,w}^{i+1}(t)$.

**Remark 4.2** In (4.3), only the local information, the estimated coordinates $\hat{\mathbf{c}}_q^i(t)$ of the follower neighbors and the real coordinates $\mathbf{c}_w^*(t)$ of the leader neighbors, are available when agent $l$ estimates its own coordinates $\hat{c}_l^{i+1}(t)$. Therefore, localization estimation algorithm (4.3) is distributed.

**Remark 4.3** The noises in the measurement side and the communication side are assumed to be zero-mean, bounded and independently distributed in Assumptions 1 and 2. Such assumptions are common for communication noise. The zero-mean assumption of the distance measurement noise is also reasonable. For example, the time to arrival (TOA) based relative distance measurements with bias can be converted into unbiased measurements [4]. In a word, Assumptions 4.1 and 4.2 are reasonable and not conservative.

For the convenience of analysis, a compact form is formulated for all $(N-m)$ followers,

$$\begin{aligned}\hat{\mathbf{C}}_s^{i+1}(t) = \left(1-\alpha^{i+1}(t)\right)\hat{\mathbf{C}}_s^i(t) + \beta^{i+1}(t)\hat{\mathbf{P}}^{i+1}(t)\times \\ \left(\hat{\mathbf{C}}_s^i(t) + \mathbf{\Phi}_s^i(t)\right) + \beta^{i+1}(t)\hat{\mathbf{B}}^{i+1}(t)\left(\mathbf{C}_a^*(t) + \mathbf{\Phi}_a^i(t)\right),\end{aligned} \tag{4.4}$$

where $\hat{\mathbf{C}}_s^i(t) = \left[\left(\hat{\mathbf{c}}_{m+1}^i(t)\right)^{\mathrm{T}}, \ldots, \left(\hat{\mathbf{c}}_N^i(t)\right)^{\mathrm{T}}\right]^{\mathrm{T}}$ denotes the estimated coordinates of all the followers, and $\mathbf{C}_a^*(t) = \left[\left(\mathbf{c}_1^*(t)\right)^{\mathrm{T}}, \ldots, \left(\mathbf{c}_m^*(t)\right)^{\mathrm{T}}\right]^{\mathrm{T}}$ represents the real coordinates of all the leaders. $\mathbf{\Phi}_s^i(t) = \left[\left(\varphi_{m+1}^i(t)\right)^{\mathrm{T}}, \ldots, \left(\varphi_N^i(t)\right)^{\mathrm{T}}\right]^{\mathrm{T}}$ and $\mathbf{\Phi}_a^i(t) = \left[\left(\varphi_1^i(t)\right)^{\mathrm{T}}, \ldots, \left(\varphi_m^i(t)\right)^{\mathrm{T}}\right]^{\mathrm{T}}$ are the communication noises between agents.

$$\hat{\mathbf{P}}^{i+1}(t) = \begin{bmatrix} 0 & \hat{a}_{m+1,m+2}^{i+1}(t) & \cdots & \hat{a}_{m+1,N}^{i+1}(t) \\ \hat{a}_{m+2,m+1}^{i+1}(t) & 0 & \cdots & \hat{a}_{m+2,N}^{i+1}(t) \\ \vdots & \vdots & \ddots & \vdots \\ \hat{a}_{N,m+1}^{i+1}(t) & \hat{a}_{N,m+2}^{i+1}(t) & \cdots & 0 \end{bmatrix} \in \mathbb{R}^{(N-m)\times(N-m)}$$ characterizes the information interaction between the followers.

$$\hat{\mathbf{B}}^{i+1}(t) = \begin{bmatrix} \hat{a}_{m+1,1}^{i+1}(t) & \hat{a}_{m+1,2}^{i+1}(t) & \cdots & \hat{a}_{m+1,m}^{i+1}(t) \\ \hat{a}_{m+2,1}^{i+1}(t) & \hat{a}_{m+2,2}^{i+1}(t) & \cdots & \hat{a}_{m+2,m}^{i+1}(t) \\ \vdots & \vdots & \ddots & \vdots \\ \hat{a}_{N,1}^{i+1}(t) & \hat{a}_{N,2}^{i+1}(t) & \cdots & \hat{a}_{N,m}^{i+1}(t) \end{bmatrix} \in \mathbb{R}^{(N-m)\times m}$$ characterizes the information interaction between the followers and the leaders.

**Remark 4.4** Matrices $\hat{\mathbf{P}}^{i+1}(t)$ and $\hat{\mathbf{B}}^{i+1}(t)$ are the estimates of $\mathbf{P}^{i+1}(t)$ and $\mathbf{B}^{i+1}(t)$, respectively. $\mathbf{P}^{i+1}(t)$ and $\mathbf{B}^{i+1}(t)$ are the nominal system's matrices composed of real barycentric coordinates.

### 4.3.2 Convergence Analysis

Before the convergence analysis, a lemma is given first.

**Lemma 4.1** ([5]) *For all iteration number k, let $u_k \geq 0$. If there exist constants $\sigma > 0$, $\delta > 0$, $0 < \lambda_1 < 1$ and $\lambda_1 < \lambda_2$ such that*

$$u_{k+1} \leq \left(1 - \frac{\sigma}{(1+k)^{\lambda_1}}\right) u_k + \frac{\delta}{(1+k)^{\lambda_2}}, \tag{4.5}$$

*then,*

$$u_k \leq \frac{\delta}{\sigma} \frac{1}{(1+k)^{\lambda_2 - \lambda_1}} + o\left(\frac{1}{(1+k)^{\lambda_2 - \lambda_1}}\right). \tag{4.6}$$

*The convergence of the localization algorithm in the presence of distance measurement noise and communication noise is provided.*

**Theorem 4.1** *Consider the dynamic multi-agent systems perform tasks repeatedly. Algorithm 4.1 solves the real-time localization problem with distance measurement and communication noises, i.e.,* $\lim_{i \to \infty} \left\| \mathbf{E}_s^{i+1}(t) \right\|_\infty = 0.$

***Proof*** Define $\tilde{\mathbf{P}}^{i+1}(t) = \hat{\mathbf{P}}^{i+1}(t) - \mathbf{P}^{i+1}(t)$, $\tilde{\mathbf{B}}^{i+1}(t) = \hat{\mathbf{B}}^{i+1}(t) - \mathbf{B}^{i+1}(t)$ represents the estimation error matrices. Then, (4.4) becomes

$$\begin{aligned}
\hat{\mathbf{C}}_s^{i+1}(t) = &\left(1 - \alpha^{i+1}(t)\right) \hat{\mathbf{C}}_s^{i}(t) + \beta^{i+1}(t)\tilde{\mathbf{P}}^{i+1}(t)\times \\
&\left(\hat{\mathbf{C}}_s^{i}(t) + \mathbf{\Phi}_s^{i}(t)\right) + \beta^{i+1}(t)\mathbf{P}^{i+1}(t)\left(\hat{\mathbf{C}}_s^{i}(t) + \mathbf{\Phi}_s^{i}(t)\right) \\
&+\beta^{i+1}(t)\tilde{\mathbf{B}}^{i+1}(t)\left(\mathbf{C}_a^{*}(t) + \mathbf{\Phi}_a^{i}(t)\right) \\
&+\beta^{i+1}(t)\mathbf{B}^{i+1}(t)\left(\mathbf{C}_a^{*}(t) + \mathbf{\Phi}_a^{i}(t)\right).
\end{aligned} \tag{4.7}$$

Define $\mathbf{E}_s^{i+1}(t) = \hat{\mathbf{C}}_s^{i+1}(t) - \mathbf{C}_s^{*}(t)$ denotes the position estimation error, where $\mathbf{C}_s^{*}(t)$ is the real coordinates of the followers. Subtracting $\mathbf{C}_s^{*}(t)$ from both sides of (4.7), one has,

$$
\begin{aligned}
&\mathbf{E}_s^{i+1}(t) = \left[1 - \alpha^{i+1}(t) + \beta^{i+1}(t)\mathbf{P}^{i+1}(t)\right]\mathbf{E}_s^{i}(t) + \\
&\beta^{i+1}(t)\tilde{\mathbf{P}}^{i+1}(t)\mathbf{E}_s^{i}(t) + \left[\left(\beta^{i+1}(t) - \alpha^{i+1}(t)\right)\mathbf{C}_s^{*}(t)\right. \\
&\left.+\beta^{i+1}(t)\left(\left(\mathbf{P}^{i+1}(t) - \mathbf{I}\right)\mathbf{C}_s^{*}(t) + \mathbf{B}^{i+1}(t)\mathbf{C}_a^{*}(t)\right)\right] \\
&+\beta^{i+1}(t)\left(\mathbf{P}^{i+1}(t)\mathbf{\Phi}_s^{i}(t) + \mathbf{B}^{i+1}(t)\mathbf{\Phi}_a^{i}(t)\right) \\
&+\beta^{i+1}(t)\tilde{\mathbf{P}}^{i+1}(t)\left(\mathbf{C}_s^{*}(t) + \mathbf{\Phi}_s^{i}(t)\right) \\
&+\beta^{i+1}(t)\tilde{\mathbf{B}}^{i+1}(t)\left(\mathbf{C}_a^{*}(t) + \mathbf{\Phi}_a^{i}(t)\right).
\end{aligned}
\tag{4.8}
$$

The fourth term on the right-hand side of (4.8) becomes

$$
\begin{aligned}
&\mathbf{B}^{i+1}(t)\mathbf{C}_a^{*}(t) + \left(\mathbf{P}^{i+1}(t) - \mathbf{I}\right)\mathbf{C}_s^{*}(t) = \\
&\begin{bmatrix}
a_{m+1,1}^{i+1}(t) & a_{m+1,2}^{i+1}(t) & \cdots & a_{m+1,m}^{i+1}(t) \\
a_{m+2,1}^{i+1}(t) & a_{m+2,2}^{i+1}(t) & \cdots & a_{m+2,m}^{i+1}(t) \\
\vdots & \vdots & \ddots & \vdots \\
a_{N,1}^{i+1}(t) & a_{N,2}^{i+1}(t) & \cdots & a_{N,m}^{i+1}(t)
\end{bmatrix}
\begin{bmatrix}
\mathbf{c}_1^{*}(t) \\ \mathbf{c}_2^{*}(t) \\ \vdots \\ \mathbf{c}_m^{*}(t)
\end{bmatrix} \\
&+\begin{bmatrix}
-1 & a_{m+1,m+2}^{i+1}(t) & \cdots & a_{m+1,N}^{i+1}(t) \\
a_{m+2,m+1}^{i+1}(t) & -1 & \cdots & a_{m+2,N}^{i+1}(t) \\
\vdots & \vdots & \ddots & \vdots \\
a_{N,m+1}^{i+1}(t) & a_{N,m+2}^{i+1}(t) & \cdots & -1
\end{bmatrix}
\begin{bmatrix}
\mathbf{c}_{m+1}^{*}(t) \\ \mathbf{c}_{m+2}^{*}(t) \\ \vdots \\ \mathbf{c}_N^{*}(t)
\end{bmatrix}.
\end{aligned}
\tag{4.9}
$$

It follows from the right side of (4.9) that

$$
\begin{gathered}
a_{m+1,1}^{i+1}(t)\mathbf{c}_1^{*}(t) + \cdots + a_{m+1,m}^{i+1}(t)\mathbf{c}_m^{*}(t) + a_{m+1,m+2}^{i+1}(t) \\
\times\mathbf{c}_{m+2}^{*}(t) + \cdots + a_{m+1,N}^{i+1}(t)\mathbf{c}_N^{*}(t) - \mathbf{c}_{m+1}^{*}(t), \\
\vdots \\
a_{N,1}^{i+1}(t)\mathbf{c}_1^{*}(t) + \cdots + a_{N,m}^{i+1}(t)\mathbf{c}_m^{*}(t) + a_{N,m+1}^{i+1}(t)\mathbf{c}_{m+1}^{*}(t) \\
\times + \cdots + a_{N,N-1}^{i+1}(t)\mathbf{c}_{N-1}^{*}(t) - \mathbf{c}_N^{*}(t).
\end{gathered}
\tag{4.10}
$$

According to $\mathbf{c}_l(t) = a_{lh}(t)\mathbf{c}_h(t) + a_{lj}(t)\mathbf{c}_j(t) + a_{lk}(t)\mathbf{c}_k(t)$, the real coordinates of the agent $l$ can be calculated when the real coordinates of the neighbors are known. One can obtain that

$$
\begin{gathered}
a_{m+1,1}^{i+1}(t)\mathbf{c}_1^{*}(t) + \cdots + a_{m+1,m}^{i+1}(t)\mathbf{c}_m^{*}(t) + a_{m+1,m+2}^{i+1}(t) \\
\times\mathbf{c}_{m+2}^{*}(t) + \cdots + a_{m+1,N}^{i+1}(t)\mathbf{c}_N^{*}(t) = \mathbf{c}_{m+1}^{*}(t), \\
\vdots \\
a_{N,1}^{i+1}(t)\mathbf{c}_1^{*}(t) + \cdots + a_{N,m}^{i+1}(t)\mathbf{c}_m^{*}(t) + a_{N,m+1}^{i+1}(t)\mathbf{c}_{m+1}^{*}(t) \\
+ \cdots + a_{N,N-1}^{i+1}(t)\mathbf{c}_{N-1}^{*}(t) = \mathbf{c}_N^{*}(t).
\end{gathered}
\tag{4.11}
$$

Thus, $\mathbf{B}^{i+1}(t)\mathbf{C}_a^{*}(t) + \left(\mathbf{P}^{i+1}(t) - \mathbf{I}\right)\mathbf{C}_s^{*}(t) = \mathbf{0}$.
Taking infinite norm on both sides of (4.8),

$$\begin{aligned}
&\left\|\mathbf{E}_s^{i+1}(t)\right\|_\infty \le \left\|1-\alpha^{i+1}(t)+\beta^{i+1}(t)\mathbf{P}^{i+1}(t)\right\|_\infty \\
&\times\left\|\mathbf{E}_s^{i}(t)\right\|_\infty + \beta^{i+1}(t)\left\|\tilde{\mathbf{P}}^{i+1}(t)\right\|_\infty\left\|\mathbf{E}_s^{i}(t)\right\|_\infty \\
&+\left(\alpha^{i+1}(t)-\beta^{i+1}(t)\right)\left\|\mathbf{C}_s^{*}(t)\right\|_\infty + \beta^{i+1}(t) \\
&\times\left(\left\|\mathbf{P}^{i+1}(t)\right\|_\infty\left\|\mathbf{\Phi}_s^{i}(t)\right\|_\infty + \left\|\mathbf{B}^{i+1}(t)\right\|_\infty\left\|\mathbf{\Phi}_a^{i}(t)\right\|_\infty\right) \\
&+\beta^{i+1}(t)\left\|\tilde{\mathbf{P}}^{i+1}(t)\right\|_\infty\left(\left\|\mathbf{C}_s^{*}(t)\right\|_\infty + \left\|\mathbf{\Phi}_s^{i}(t)\right\|_\infty\right) \\
&+\beta^{i+1}(t)\left\|\tilde{\mathbf{B}}^{i+1}(t)\right\|_\infty\left(\left\|\mathbf{C}_a^{*}(t)\right\|_\infty + \left\|\mathbf{\Phi}_a^{i}(t)\right\|_\infty\right).
\end{aligned} \tag{4.12}$$

Since the element in matrix $\mathbf{P}_{i+1}(t)$ represents the barycentric coordinates between follower agents, one can guarantee that the sum of each row of matrix $\mathbf{P}_{i+1}(t)$ is less than 1. According to the property of infinite norm, one has $\left\|\mathbf{P}^{i+1}(t)\right\|_\infty < 1$.

Due to the fact that $\alpha^{i+1}(t) > \beta^{i+1}(t) > 0$, the first term on the right-hand side of (4.12) becomes

$$\begin{aligned}
&\left\|1-\alpha^{i+1}(t)+\beta^{i+1}(t)\mathbf{P}^{i+1}(t)\right\|_\infty\left\|\mathbf{E}_s^{i}(t)\right\|_\infty \\
&\le\left(1-\alpha^{i+1}(t)+\beta^{i+1}(t)\left\|\mathbf{P}^{i+1}(t)\right\|_\infty\right)\left\|\mathbf{E}_s^{i}(t)\right\|_\infty \\
&\le\left(1-\alpha^{i+1}(t)+\alpha^{i+1}(t)\left\|\mathbf{P}^{i+1}(t)\right\|_\infty\right)\left\|\mathbf{E}_s^{i}(t)\right\|_\infty \\
&=1-\alpha^{i+1}(t)\left(1-\left\|\mathbf{P}^{i+1}(t)\right\|_\infty\right)\left\|\mathbf{E}_s^{i}(t)\right\|_\infty \\
&\le\left(1-\alpha^{i+1}(t)\delta_1\right)\left\|\mathbf{E}_s^{i}(t)\right\|_\infty,
\end{aligned} \tag{4.13}$$

where $\delta_1 = 1-\left\|\mathbf{P}^{i+1}(t)\right\|_\infty$.

Similarly, one can obtain that

$$\beta^{i+1}(t)\left\|\tilde{\mathbf{P}}^{i+1}(t)\right\|_\infty\left\|\mathbf{E}_s^{i}(t)\right\|_\infty \le \alpha^{i+1}(t)\left\|\tilde{\mathbf{P}}^{i+1}(t)\right\|_\infty\left\|\mathbf{E}_s^{i}(t)\right\|_\infty. \tag{4.14}$$

There exists an iteration number $i_b$ sufficient largely such that, for $i \ge i_b$, $\left\|\tilde{\mathbf{P}}^{i+1}(t)\right\|_\infty \le \xi$, $\left\|\tilde{\mathbf{B}}^{i+1}(t)\right\|_\infty \le \xi$, where $\xi < 1-\left\|\mathbf{P}^{i+1}(t)\right\|_\infty$.

**Remark 4.5** The estimated relative distance $\hat{d}_{xy}^{i}(t)$ will converge to the real relative distance $d_{xy}^{i}(t)$ with the increase of the iterations, so the calculated barycentric coordinates $\hat{a}_{l,q}^{i+1}(t)$ and $\hat{a}_{l,w}^{i+1}(t)$ will converge to the real barycentric coordinates $a_{l,q}^{i+1}(t)$ and $a_{l,w}^{i+1}(t)$. In other words, according to the strong law of large numbers, one can obtain that $\lim_{i\to\infty}\hat{\mathbf{P}}^{i+1}(t) = \mathbf{P}^{i+1}(t)$. Therefore, there exists an iteration number sufficient largely such that $\left\|\tilde{\mathbf{P}}^{i+1}(t)\right\|_\infty$ will converge to a small bounded range, and $\left\|\mathbf{P}^{i+1}(t)\right\|_\infty + \left\|\tilde{\mathbf{P}}^{i+1}(t)\right\|_\infty < 1$ holds.

Then, one has

$$
\begin{aligned}
&\left\|1-\alpha^{i+1}(t)+\beta^{i+1}(t)\mathbf{P}^{i+1}(t)\right\|_{\infty}\left\|\mathbf{E}_s^i(t)\right\|_{\infty} \\
&+\beta^{i+1}(t)\left\|\tilde{\mathbf{P}}^{i+1}(t)\right\|_{\infty}\left\|\mathbf{E}_s^i(t)\right\|_{\infty} \\
&\le \left(1-\alpha^{i+1}(t)\delta_1\right)\left\|\mathbf{E}_s^i(t)\right\|_{\infty}+\alpha^{i+1}(t)\xi\left\|\mathbf{E}_s^i(t)\right\|_{\infty} \\
&=\left(1-\alpha^{i+1}(t)\left(\delta_1-\xi\right)\right)\left\|\mathbf{E}_s^i(t)\right\|_{\infty}.
\end{aligned} \tag{4.15}
$$

Since $\lambda_1 < \lambda_2$ and $\lambda_2 - \lambda_1 < \zeta$. There exists an iteration number $i_o$ sufficient largely such that, for $i \ge i_o$, $\beta^{i+1}(t) \to \alpha^{i+1}(t)$. Therefore, when $i \ge i_o$, the third term $\left(\alpha^{i+1}(t) - \beta^{i+1}(t)\right)\left\|\mathbf{C}_s^*(t)\right\|_{\infty}$ on the right-hand side of (4.12) is negligible for the bounded $\left\|\mathbf{C}_s^*(t)\right\|_{\infty}$.

According to the Assumption 4.2, one can know that the communication noises $\left\|\mathbf{\Phi}_s^i(t)\right\|_{\infty}$ and $\left\|\mathbf{\Phi}_a^i(t)\right\|_{\infty}$ is bounded. Since the element in matrix $\mathbf{B}^{i+1}(t)$ represents the barycentric coordinate between the follower agents and the leader agents, $\left\|\mathbf{B}^{i+1}(t)\right\|_{\infty}$ is bounded. $\mathbf{C}_s^*(t)$ and $\mathbf{C}_a^*(t)$ represent the real coordinates, $\left\|\mathbf{C}_s^*(t)\right\|_{\infty}$ and $\left\|\mathbf{C}_a^*(t)\right\|_{\infty}$ are bounded. The last three terms on the right-hand side of (4.12) can be rewritten as

$$
\begin{aligned}
&\beta^{i+1}(t)\left(\left\|\mathbf{P}^{i+1}(t)\right\|_{\infty}\left\|\mathbf{\Phi}_s^i(t)\right\|_{\infty}+\left\|\mathbf{B}^{i+1}(t)\right\|_{\infty}\left\|\mathbf{\Phi}_a^i(t)\right\|_{\infty}\right) \\
&+\beta^{i+1}(t)\left\|\tilde{\mathbf{P}}^{i+1}(t)\right\|_{\infty}\left(\left\|\mathbf{C}_s^*(t)\right\|_{\infty}+\left\|\mathbf{\Phi}_s^i(t)\right\|_{\infty}\right) \\
&+\beta^{i+1}(t)\left\|\tilde{\mathbf{B}}^{i+1}(t)\right\|_{\infty}\left(\left\|\mathbf{C}_a^*(t)\right\|_{\infty}+\left\|\mathbf{\Phi}_a^i(t)\right\|_{\infty}\right) \\
&\le \beta^{i+1}(t)\delta_2,
\end{aligned} \tag{4.16}
$$

where $\delta_2 < \infty$ is bounded.

Let $i_{\max} = \max\left(i_b, i_o\right)$, it follows from (4.13) to (4.16), for $i \ge i_{\max}$, (4.12) becomes

$$
\begin{aligned}
\left\|\mathbf{E}_s^{i+1}(t)\right\|_{\infty} &\le \left(1-\alpha^{i+1}(t)\left(\delta_1-\xi\right)\right)\left\|\mathbf{E}_s^i(t)\right\|_{\infty}+\beta^{i+1}(t)\delta_2 \\
&=\left(1-\alpha^{i+1}(t)\delta_3\right)\left\|\mathbf{E}_s^i(t)\right\|_{\infty}+\beta^{i+1}(t)\delta_2,
\end{aligned} \tag{4.17}
$$

where $\delta_3 = \delta_1 - \xi$.

According to Lemma 4.1, one can obtain that

$$
\left\|\mathbf{E}_s^{i+1}(t)\right\|_{\infty} \le \frac{\delta_2}{\delta_3}\frac{1}{(1+i)^{\lambda_2-\lambda_1}}+o\left(\frac{1}{(1+i)^{\lambda_2-\lambda_1}}\right). \tag{4.18}
$$

Since $\frac{\delta_2}{\delta_3}$ is bounded, (4.18) implies $\lim\limits_{i\to\infty}\left\|\mathbf{E}_s^{i+1}(t)\right\|_{\infty} = 0$.

The proof of Theorem 4.1 is completed.

## 4.4 Distributed Localization Algorithm with Measurement Noise

In this section, the localizability problem of the multi-agent systems with only measurement noise is discussed as a corollary. The objective is to design a concise and effective distributed localization estimation algorithm to eliminate the effect of the measurement noise on the localization accuracy.

### 4.4.1 Estimator Design

A distributed localization estimation algorithm based on iterative learning is proposed as follows

$$\hat{\mathbf{c}}_l^{i+1}(t) = \sum_{q \in N_l} \hat{a}_{l,q}^{i+1}(t)\hat{\mathbf{c}}_q^i(t) + \sum_{w \in N_l} \hat{a}_{l,w}^{i+1}(t)\mathbf{c}_w^*(t), \tag{4.19}$$

The process of distributed localization estimation algorithm based on iterative learning is presented in Algorithm 4.2.

---

**Algorithm 4.2.** Estimate the real coordinate of the follower agent with distance measurement noise

---

**for** $i = 1 : Iteration.$
**initialization** Given the initial iteration estimated coordinates $\hat{\mathbf{c}}_l^i(t)$, $l \in \{m+1, \ldots, N\}$.
**Step 1.** Each agent sends its own estimated coordinates to its neighbors.
**Step 2.** Measure the relative distance $\tilde{d}_{xy}^i(t)$ and calculate the estimated relative distance $\hat{d}_{xy}^i(t)$:

$$\hat{d}_{xy}^i(t) = \frac{1}{i}\sum_{n=0}^{i} \tilde{d}_{xy}^n(t) = \frac{i-1}{i}\hat{d}_{xy}^{i-1}(t) + \frac{1}{i}\tilde{d}_{xy}^i(t).$$

**Step 3.** Calculate the barycentric coordinates $\hat{a}_{l,q}^{i+1}(t)$ and $\hat{a}_{l,w}^{i+1}(t)$ according to Cayley-Menger determinant.
**Step 4.** Update $\hat{\mathbf{c}}_l^{i+1}(t) = \sum_{q \in N_l} \hat{a}_{l,q}^{i+1}(t)\hat{\mathbf{c}}_q^i(t) + \sum_{w \in N_l} \hat{a}_{l,w}^{i+1}(t)\mathbf{c}_w^*(t)$.
**Step 5.** Update $i = i + 1$ and return to **Step 1**.
**end**

---

### 4.4.2 Convergence Analysis

The convergence of the localization algorithm in the presence of the measurement noise is provided.

**Theorem 4.2** *Consider the dynamic multi-agent systems perform tasks repeatedly. Algorithm 4.2 solves the real-time localization problem with distance measurement noise, i.e.,* $\lim_{i\to\infty}\left\|\mathbf{E}_s^{i+1}(t)\right\|_\infty = 0$.

***Proof*** Consider all $(N-m)$ followers, one has

$$\hat{\mathbf{C}}_s^{i+1}(t) = \hat{\mathbf{P}}^{i+1}(t)\hat{\mathbf{C}}_s^{i}(t) + \hat{\mathbf{B}}^{i+1}(t)\mathbf{C}_a^*(t), \tag{4.20}$$

According to the definition of $\tilde{\mathbf{P}}^{i+1}(t)$ and $\tilde{\mathbf{B}}^{i+1}(t)$, (4.20) becomes

$$\begin{aligned}\hat{\mathbf{C}}_s^{i+1}(t) = {}& \tilde{\mathbf{P}}^{i+1}(t)\hat{\mathbf{C}}_s^{i}(t) + \tilde{\mathbf{B}}^{i+1}(t)\mathbf{C}_a^*(t)\\ &+\mathbf{P}^{i+1}(t)\hat{\mathbf{C}}_s^{i}(t) + \mathbf{B}^{i+1}(t)\mathbf{C}_a^*(t).\end{aligned} \tag{4.21}$$

Subtracting $\mathbf{C}_s^*(t)$ from both sides of (4.21), one can obtain that

$$\begin{aligned}\mathbf{E}_s^{i+1}(t) = {}& \mathbf{P}^{i+1}(t)\mathbf{E}_s^{i}(t) + \tilde{\mathbf{P}}^{i+1}(t)\mathbf{E}_s^{i}(t) +\\ &\left[\left(\mathbf{P}^{i+1}(t)-\mathbf{I}\right)\mathbf{C}_s^*(t) + \mathbf{B}^{i+1}(t)\mathbf{C}_a^*(t)\right]\\ &+\tilde{\mathbf{P}}^{i+1}(t)\mathbf{C}_s^*(t) + \tilde{\mathbf{B}}^{i+1}(t)\mathbf{C}_a^*(t).\end{aligned} \tag{4.22}$$

Since $\mathbf{B}^{i+1}(t)\mathbf{C}_a^*(t) + \left(\mathbf{P}^{i+1}(t)-\mathbf{I}\right)\mathbf{C}_s^*(t) = \mathbf{0}$, one has

$$\begin{aligned}\mathbf{E}_s^{i+1}(t) = {}& \mathbf{P}^{i+1}(t)\mathbf{E}_s^{i}(t) + \tilde{\mathbf{P}}^{i+1}(t)\mathbf{E}_s^{i}(t)\\ &+\tilde{\mathbf{P}}^{i+1}(t)\mathbf{C}_s^*(t) + \tilde{\mathbf{B}}^{i+1}(t)\mathbf{C}_a^*(t).\end{aligned} \tag{4.23}$$

Taking infinite norms on both sides of (4.23), one has

$$\begin{aligned}\left\|\mathbf{E}_s^{i+1}(t)\right\|_\infty \le {}& \left(\left\|\mathbf{P}^{i+1}(t)\right\|_\infty + \left\|\tilde{\mathbf{P}}^{i+1}(t)\right\|_\infty\right)\left\|\mathbf{E}_s^{i}(t)\right\|_\infty\\ &+\left\|\tilde{\mathbf{P}}^{i+1}(t)\right\|_\infty\left\|\mathbf{C}_s^*(t)\right\|_\infty + \left\|\tilde{\mathbf{B}}^{i+1}(t)\right\|_\infty\left\|\mathbf{C}_a^*(t)\right\|_\infty.\end{aligned} \tag{4.24}$$

Refer to Remark 4.5, one can obtain that

$$\left\|\mathbf{P}^{i+1}(t)\right\|_\infty + \left\|\tilde{\mathbf{P}}^{i+1}(t)\right\|_\infty = \sigma_1 < 1. \tag{4.25}$$

Therefore, (4.24) becomes

$$\left\|\mathbf{E}_s^{i+1}(t)\right\|_\infty \le \sigma_1\left\|\mathbf{E}_s^{i}(t)\right\|_\infty + \xi\sigma_2, \tag{4.26}$$

where $\sigma_2 = \left\|\mathbf{C}_s^*(t)\right\|_\infty + \left\|\mathbf{C}_a^*(t)\right\|_\infty < \infty$ is bounded.

By solving (4.26), one can obtain that

$$\left\|\mathbf{E}_s^{i+1}(t)\right\|_\infty \leq \sigma_1^{i-i_b+1}\left\|\mathbf{E}_s^{i_b}(t)\right\|_\infty + \xi\frac{\sigma_2}{1-\sigma_1}. \tag{4.27}$$

Since $0 < \sigma_1 < 1$, then, one has

$$\lim_{i\to\infty} \sigma_1^{i-i_b+1}\left\|\mathbf{E}_s^{i_b}(t)\right\|_\infty = 0. \tag{4.28}$$

Similarly, one can obtain that $\frac{\sigma_2}{1-\sigma_1} = \sigma_3 < \infty$ is bounded. Note that, $\sigma_3$ and $\xi$ are independent. We then have from (4.27) that

$$\limsup_{i\to\infty} \left\|\mathbf{E}_s^{i+1}(t)\right\|_\infty \leq \xi\sigma_3. \tag{4.29}$$

Since (4.29) holds for arbitrary $\xi$, then (4.29) implies that $\lim\limits_{i\to\infty}\left\|\mathbf{E}_s^{i+1}(t)\right\|_\infty = 0$. The proof of Theorem 4.2 is completed.

## 4.5 Simulation and Experiment

In this section, the numerical simulation and the robot experiment are provided to testify the effectiveness and realizability of the proposed algorithms.

### *4.5.1 Numerical Simulation*

Seven agents are considered, which move along an elliptical trajectory in two-dimensional plane. The directed topology between agents is shown in Fig. 4.1, where the follower agents are located in the convex hull formed by the leaders. The real trajectory curves of all agents are shown in Fig. 4.2, where the asterisk represents the initial position of the agent and the arrow represents the move direction of the agent. In order to save space, this subsection will discuss the convergence of the Algorithm 4.1 with measurement and communication noises.

The distance measurement noise $\upsilon_{xy}^i(t)$ obeys Gaussian distribution, with zero-mean and the variance $\frac{d_{xy}(t)}{10}$ along the iteration axis. The communication noise varying randomly in $[-0.5, 0.5]$ is imposed on the estimated coordinates.

The initial estimated values of all followers are given randomly. Select $\lambda_1 = 0.80$ and $\lambda_2 = 0.83$. Applying Algorithm 4.1, the maximum estimation errors $e_{l,\max}(i) = \max\limits_t \left\{\left\|\hat{\mathbf{c}}_l^i(t) - \mathbf{c}_l^*(t)\right\|_2\right\}$ along the iteration axis are shown in Fig. 4.3. It can be seen that the maximum estimation errors can asymptotically converge to zero. Meanwhile, for comparison purposes, the maximum estimation error curves under ideal

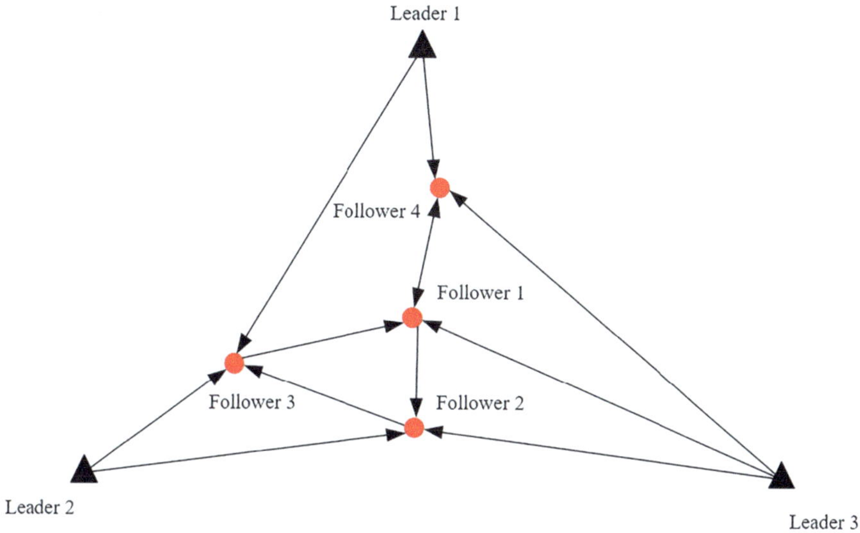


**Fig. 4.1** The directed topology of seven agents

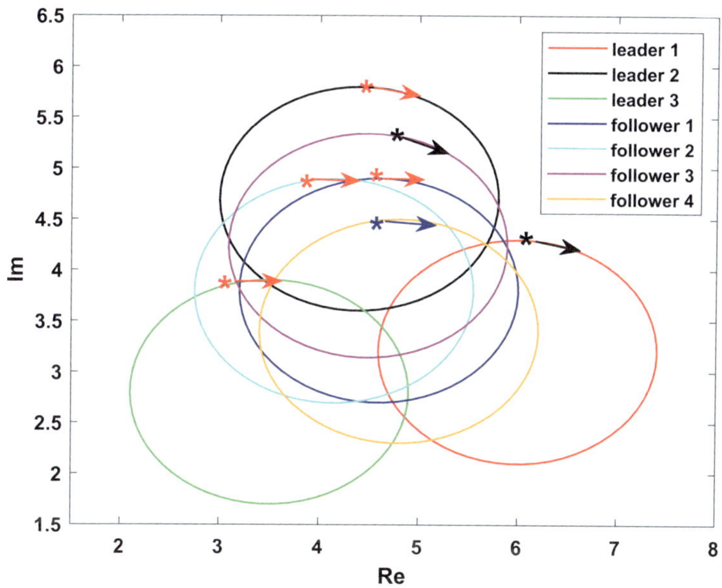


**Fig. 4.2** The trajectories of seven agents

environment (no measurement and communication noise) are shown in Fig. 4.4. Furthermore, the estimation error curves in the case of only measurement noise is shown in Fig. 4.5.

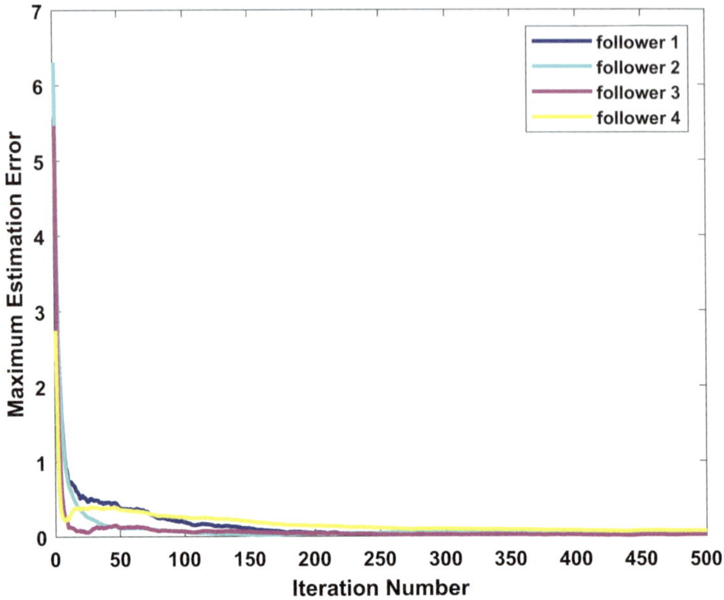


**Fig. 4.3** The maximum estimation error under bounded measurement and communication noises

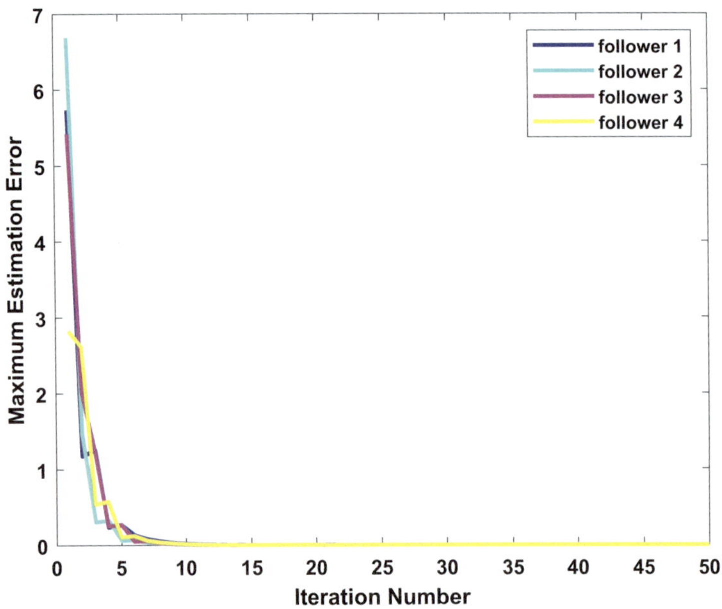


**Fig. 4.4** The maximum estimation error under ideal environment

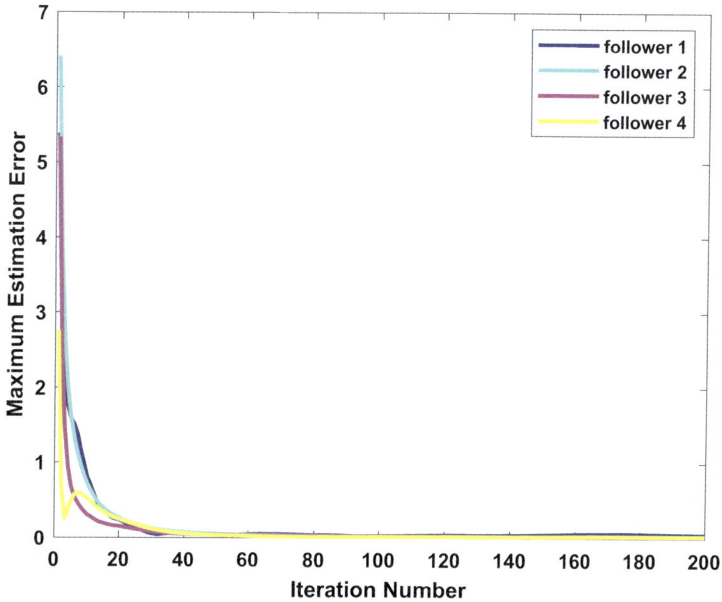


**Fig. 4.5** The maximum estimation error under bounded measurement noises

Comparing with Figs. 4.3, 4.4 and 4.5, one can find that although the convergence speed of the estimation error in the case of measurement and communication noises is slower than that in the case of ideal environment, Algorithm 4.1 with unbiased estimator and stochastic approximation method can guarantee the smooth and gradual convergence of the estimation error.

Through the simulation study, one can draw a conclusion that the proposed estimation methods can realize the real-time accurate localization of the multi-agent systems with measurement and communication noises.

## *4.5.2 QBot-2e Robot Experiment*

Five QBot-2e robots are used to test the validity of the proposed methods in this subsection. The QBot-2e experimental platform is shown in Fig. 4.6. The directed topology between five robots is shown in Fig. 4.7, where the three robots in outer circle are leaders and the two robots in inner circle are followers. The leader robots know their real coordinates, the follower robots estimate their coordinates through the proposed estimation methods. Similar to the Sect. 4.5.1, the convergence of the Algorithm 4.1 with measurement and communication noises is discussed.

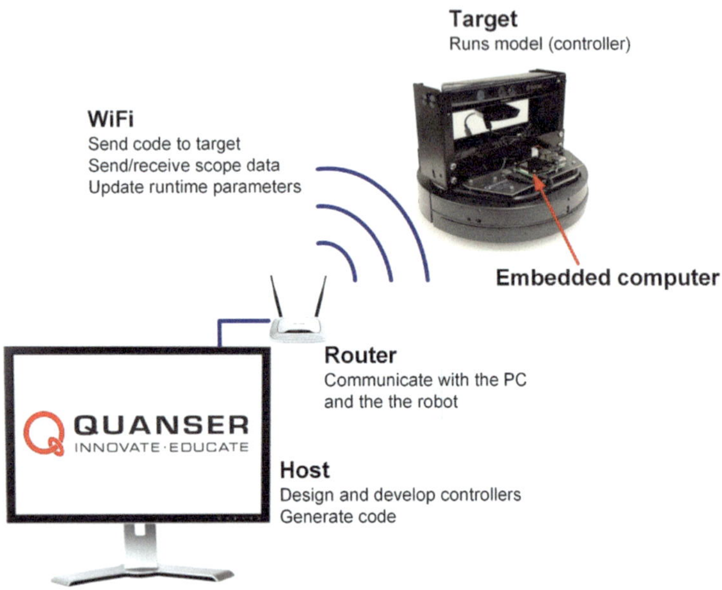


**Fig. 4.6** QBot-2e experimental platform

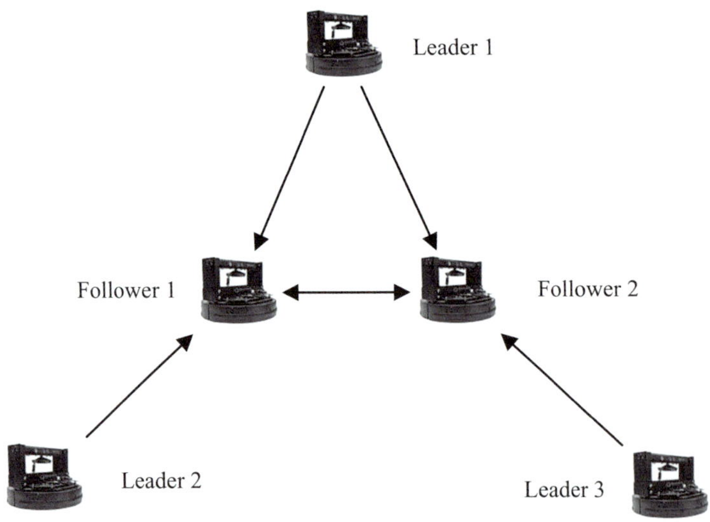


**Fig. 4.7** The directed graph of five robots

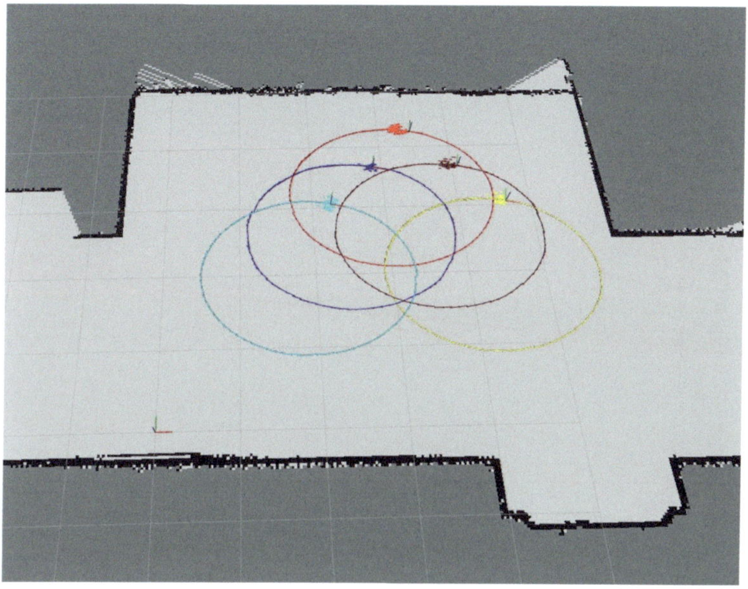

**Fig. 4.8** The real trajectory curves of five robots

The real trajectory curves of the robots in experiment are shown in Fig. 4.8, where one grid is 0.6 m. In the experiment, the real coordinate of the robot is obtained by self-built map and cameras. The velocity of the robot is 0.4 m/s. The Laser Lidar is used to measure the relative distance between robots and its sampling frequency is 60 Hz.

Set the communication noise varying randomly in $[-0.5, 0.5]$. Since there is measurement noise when the Laser Lidar measures the relative distance between robots, no additional measurement noise is added in this subsection. The initial estimated values of two follower robots are given randomly. Select $\lambda_1 = 0.75$ and $\lambda_2 = 0.81$. Applying Algorithm 4.1, the maximum estimation error $e_{l,\max}(i) = \max_t \left\{ \left\| \hat{\mathbf{c}}_l^i(t) - \mathbf{c}_l^*(t) \right\|_2 \right\}$ along the iteration axis are shown in Fig. 4.9, where the unit is meter.

Figure 4.9 are the maximum estimation error curves of the two follower robots. One can find that the convergence of the estimation error curves is smooth and asymptotically. The estimation error converges to a small bounded value at the 500th iteration, and tends to converge to zero.

The effectiveness and realizability of the proposed algorithms are verified by the simulation and experimental results. One can draw the conclusion that the proposed algorithms can solve the real-time localization problem of the multi-agent systems with measurement and communication noises under directed graph.

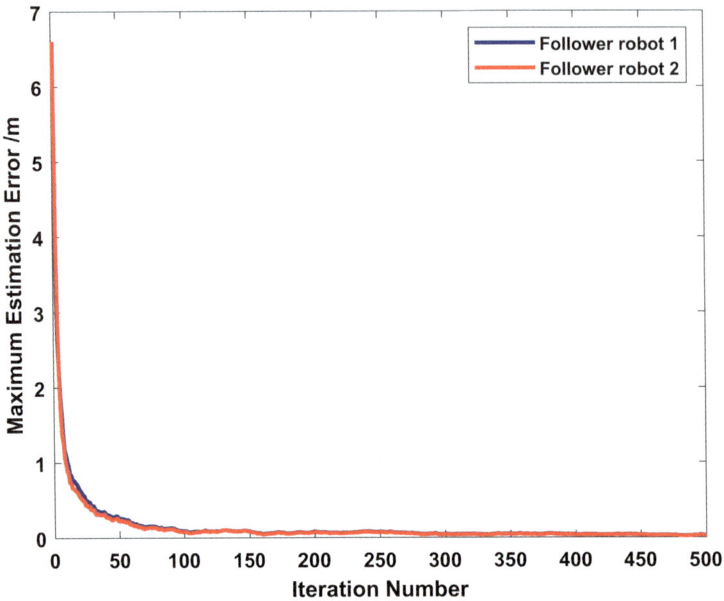


**Fig. 4.9** The maximum estimation error under bounded measurement noise and communication noise

## 4.6 Conclusions

This chapter investigated the real-time localization problem for the dynamic multi-agent systems with measurement and communication noises under directed graphs. A robust distributed localization estimation algorithm based on iterative learning is proposed. The constructed relative-distance unbiased estimator based on the historical iterative information suppresses the measurement noise. The designed stochastic approximation method with iterative-varying gains inhibits the communication noise. A rigorous asymptotic convergence analysis of the proposed algorithms is derived. The effectiveness and the practicability of the proposed methods is verified through the numerical simulation and the QBot-2e robot experiment. The performance of the localization algorithm in measurement and communication noisy environments is studied in this chapter. The localization of the multi-agent systems with arbitrary deployment in three-dimensional space under random environments is our future research interest.

# References

1. Khan, U., Kar, S., Moura, J.: DILAND: an algorithm for distributed sensor localization with noisy distance measurements. IEEE Trans. Signal Process. **58**(3), 1940–1947 (2010)
2. Wang, B., Tian, Y.: Distributed network localization: accurate estimation with noisy measurement and communication information. IEEE Trans. Signal Process. **66**(2), 5927–5940 (2018)
3. Prohorov, U.: On the strong law of large numbers. Dokl. Akad. Nauk SSSR **69**, 607–610 (1949)
4. Patwari, N.: Location Estimation in Sensor Networks. Ph.D. Dissertation, University of Michigan (2005)
5. Polyak, B.: Introduction to optimization. Optim. Softw. (1987)

# Chapter 5
# Distributed Localization of MASs with Randomly Varying Trajectory Lengths

This chapter considered the persistent real-time localization problem of dynamic multi-agent systems with repetitive running characteristics under directed graph. The trajectory length of the agent is randomly varying caused by system constraints and external environment. A novel distributed iterative learning localization estimation method with full historical average data compensation is designed. The average value of all available historical data in the previous operations is used to compensate the incomplete trajectory. The designed diagnosis mechanism and search mechanism are used to determine whether the agent has stopped running and to screen out the available historical data. In order to reduce the computation burden, an improved distributed localization algorithm with receding horizon historical average data compensation is proposed. The asymptotic convergence of the estimation algorithms in the sense of mathematical expectation is derived through the rigorous analysis. Meanwhile, the influence of the incomplete repetitive trajectory of the agent on estimation error and convergence rate is also analyzed. The experimental result based on QBot-2e robot platform verifies the realizability of the proposed method.

## 5.1 Introduction

In the process of position estimation, it is expected that the agent can keep a complete trajectory during operations. However, in reality, due to the interference of external environment or hardware failure, the agent may stop running ahead of time, then the trajectory data will be incomplete. Thus, the estimator cannot finish the normal localization estimation, which will affect the accuracy and convergence rate. Only the cooperative control of the multi-agent systems based on iterative learning under varying trajectory lengths have been reported in recent years [1, 2]. Therefore, it is of great significance to guarantee the real-time localization of dynamic multi-agent systems under randomly varying trajectory. The influence of the trajectory deviation

 
Y. Lv et al., *Distributed Localization and Formation Control of Multi-Agent Systems with Local Sensing Information*, https://doi.org/10.1007/978-981-96-6015-5_5

caused by the incomplete repeated operation on the localization accuracy should not be ignored.

In this chapter, the problem of persistent real-time accurate localization of dynamic multi-agent systems with repetitive running characteristics and randomly varying trajectory lengths under directed graph is considered. A novel distributed iterative learning localization estimation method with full historical average data compensation is proposed. Before designing the estimation algorithm, a diagnosis mechanism and a search mechanism are first proposed, which aims to judge whether the agent has stopped running and screen out all available historical information. The barycentric coordinates is introduced to formulate the relative position relationship between agents. The Bernoulli random variable is used to represent the random trajectory length of the agent and the varying trajectory length of each agent is independent. If the diagnosis mechanism judges that the agent is running normally, the algorithm without compensation measures is used to estimate the position. If the agent stops running unexpectedly, the barycentric coordinates will not be calculated normally. The historical data selected by the search mechanism will be used to guarantee a persistent localization. The result also extends to the algorithm with receding horizon historical average data compensation to reduce the computation burden. The asymptotic convergence in the sense of mathematical expectation is derived through the rigorous analysis. Meanwhile, the influence of the incomplete repetitive trajectory of the agent on the estimation error and convergence rate is also analyzed. The numerical simulation and the QBot-2e robot experiment verify the effectiveness and realizability of the proposed methods.

The contribution of this work lies in:

(i) A novel distributed iterative learning estimation method with compensation measures is proposed, which can guarantee a persistent localization.
(ii) The accurate real-time localization of dynamic multi-agent systems with randomly varying trajectory length is realized.
(iii) The real-time localization problem of dynamic multi-agent systems with fast changing characteristic is investigated under directed graph.

The rest of this chapter is arranged as follows: Sect. 5.2 shows the preliminaries and the problem descriptions. The estimator design and convergence analysis are introduced in Sect. 5.3. Section 5.4 extends the result to distributed localization estimation with receding horizon compensation. In Sect. 5.5, the numerical simulation and the robot experiment are used to verify the validity of the method. Finally, Sect. 5.6 is the conclusion.

## 5.2 Problem Formulation

Consider dynamic multi-agent systems with $m$ leaders and $(N - m)$ followers. Information transfer between agents is described by a directed graph. The leader can obtain the coordinates through its own equipped localization device, and the follower estimates coordinates through indirectly or directly communication with leaders.

In the process of localization, we expect the multi-agent systems with repetitive running characteristics could always keep a complete trajectory length. However, affected by the uncertainty of the external environment and hardware problems, the agent may stop running in advance, and cannot finish a complete task or cannot maintain a complete trajectory length. Therefore, the actual trajectory length will vary among different iterations. The relative distance between agents cannot be measured normally, and then the barycentric coordinates cannot be obtained. In this case, define the time instant $t \in \{0, 1, \ldots, \Gamma^i\}$, $\Gamma^i$ is the actual trajectory length of the $i$th iteration, where $i$ represents the iteration number.

In addition, it is reasonable to have a lower bound $\Gamma_{\min}$ for actual trajectory length of agent. For example, as shown in Fig. 5.1, for sampling time $t = \Gamma_{\min}$, the agent trajectory always exists. If $t > \Gamma_{\min}$, the agent trajectory length will vary randomly with some probability along the iteration axis.

The objective of this chapter is to design an effective distributed localization algorithm for dynamic multi-agent systems with randomly varying trajectory length. The proposed scheme can guarantee the estimation error $\mathbf{e}(t)$ converges to a small bounded value even the trajectory length varying between different iterations and the trajectory of the agent is not completely repeated, where $\mathbf{e}(t) = \hat{\mathbf{c}}(t) - \mathbf{c}^*(t)$, $\hat{\mathbf{c}}(t)$ represents the estimation coordinates and $\mathbf{c}^*(t)$ is the real coordinates.

## 5.3 Distributed Localization Estimation with Full Historical Average Data Compensation

Consider the case that all followers are located in convex hull formed by their neighbors. And the trajectory length of the agent varies randomly along the iteration axis, an example in Fig. 5.1. Without loss of generality, the localization problem in two-dimensional space is considered. The objective in this section is to develop

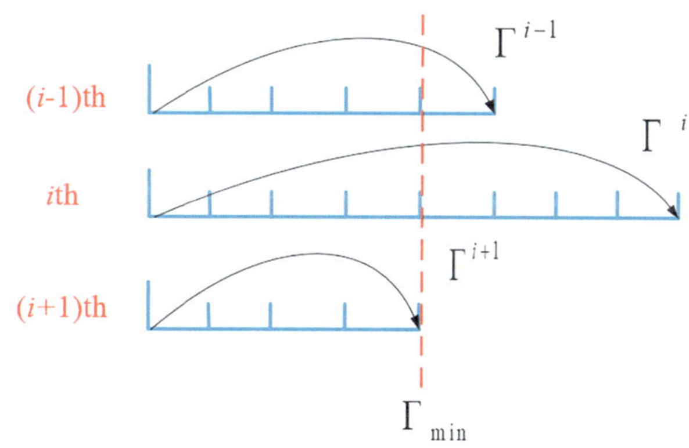


**Fig. 5.1** The varying trajectory length among different iterations

a distributed localization estimation algorithm with compensation to guarantee the accurate and real-time localization of dynamic multi-agent systems.

Note that, unlike localization algorithm in existing literature, the algorithm to be designed not only consider the persistent real-time localization problem under varying trajectory length, but also the localization accuracy problem under the trajectory of the agent is not completely repeated.

### 5.3.1 Design of Diagnostic Mechanism

When agent $l$ stops running unexpectedly at the $(t-1)$th sampling time due to the influence of external environment or its own hardware failure, its neighbors still use the equipped sensors to measure the relative distance $d(t)$ at the $t$th sampling time. However, the barycentric coordinate value $a(t)$ calculated based on the abnormal relative distance $d(t)$ at this sampling time is incorrect, which will affect the final localization accuracy. Therefore, we should first design a fault detection mechanism to determine whether the agent has stopped running at the $(t-1)$th sampling time. Before designing the diagnosis mechanism, an assumption is given below.

**Assumption 5.1** For a sampling time instant $t \in \left\{0, 1, \ldots, \Gamma^i\right\}$ in $i$th iteration, there exists at least one iteration whose length $\Gamma^\kappa, \kappa \in \{1, \ldots, i-1\}$ is greater than $t$ in previous iterations.

Select a barycentric coordinate $a^\kappa(t), \kappa \in \{1, \ldots, i-1\}$ as the standard value from previous iterations, and set a reasonable percentage value $\varphi$ according to the actual application scene, such as the velocity, the sampling interval $\Delta$ and so on. After calculating the barycentric coordinates $a^i(t)$, one can make a simple diagnosis. If $\frac{\left|a^i(t)-a^\kappa(t)\right|}{a^\kappa(t)} \leq \varphi$, then we can conclude that the agent runs normally. Otherwise, it indicates that the barycentric coordinate value $a^i(t)$ is abnormal and it should be discarded. Then the corresponding compensation measures should be taken.

### 5.3.2 Design of Search Mechanism

Before designing the compensation algorithm, we should consider how to find useful historical data. For agent $l$, a search mechanism is designed as follows.

Let $\Gamma_q^{i+1}$ is the actual trajectory length of the follower neighbor agents in the $(i+1)$th iteration, and $\Gamma_w^{i+1}$ represents the actual trajectory length of the leader neighbor agents in the $(i+1)$th iteration. Define two sets: $\Lambda_q^{i+1} = \left\{\Gamma_q^j \,\middle|\, t \leq \Gamma_q^j, j = 1, \ldots, i\right\}$ and $\Lambda_w^{i+1} = \left\{\Gamma_w^j \,\middle|\, t \leq \Gamma_w^j, j = 1, \ldots, i\right\}$. $n_q^{i+1}$ and $n_w^{i+1}$ represent the number of elements in $\Lambda_q^{i+1}$ and $\Lambda_w^{i+1}$, respectively.

**Remark 5.1** The purpose of setting $\Lambda_q^{i+1}$ and $\Lambda_w^{i+1}$ is to find out all the available barycentric coordinates calculated from the initial iteration to the latest iteration, so as to prepare for the design of the subsequent compensation algorithm.

### *5.3.3 Design of Compensation Mechanism*

A novel full historical average data compensation based distributed localization estimation algorithm is proposed,

$$\begin{aligned}\hat{\mathbf{c}}_l^{i+1}(t) = &\left(\sum_{q\in N_l}\left(\delta(1-\beta_q^{i+1}(t))\frac{1}{n_q^{i+1}}\sum_{\tau=0}^{i} a_{l,q}^{\tau}(t)\right)\hat{\mathbf{c}}_q^{i}(t)\right.\\ &\left.+\sum_{w\in N_l}\left(\delta(1-\beta_w^{i+1}(t))\frac{1}{n_w^{i+1}}\sum_{\tau=0}^{i} a_{l,w}^{\tau}(t)\right)\mathbf{c}_w^{i}(t)\right)\\ &+\sum_{q\in N_l}\beta_q^{i+1}(t)a_{l,q}^{i+1}(t)\hat{\mathbf{c}}_q^{i}(t)+\sum_{w\in N_l}\beta_w^{i+1}(t)a_{l,w}^{i+1}(t)\mathbf{c}_w^{i}(t),\end{aligned} \tag{5.1}$$

where $N_l$ represents the neighboring agent of $l$, $\beta_q^{i+1}(t)$ and $\beta_w^{i+1}(t)$ are two Bernoulli random variables, which equal to 1 or 0 indicate that the neighboring agents still running or stop running at the $t$th sampling time of the $(i+1)$th iteration; $0<\delta<1$ is a constant coefficient; $\mathbf{c}_w^{i}(t)$ and $\hat{\mathbf{c}}_q^{i}(t)$ represent the coordinate values of its neighboring agents. $\frac{1}{n_q^{i+1}}\sum_{\tau=0}^{i} a_{l,q}^{\tau}(t)$ and $\frac{1}{n_w^{i+1}}\sum_{\tau=0}^{i} a_{l,w}^{\tau}(t)$ are two iterative averaging operators to deal with the historical data.

The probability of $\beta_q^{i+1}(t)$ and $\beta_w^{i+1}(t)$ are defined as, $prob\left\{\beta_q^{i+1}(t)=1\right\}=\mathbb{E}\left\{\beta_q^{i+1}(t)\right\}=\bar{\beta}_q(t),\ \ 0<\bar{\beta}_q(t)<1$ and $prob\left\{\beta_w^{i+1}(t)=1\right\}=\mathbb{E}\left\{\beta_w^{i+1}(t)\right\}=\bar{\beta}_w(t), 0<\bar{\beta}_w(t)<1$, where $\mathbb{E}\{\}$ is the mathematical expectation.

As long as one neighbor agent of $l$ stops running, the barycentric coordinates required in the estimation algorithm cannot be calculated normally. Therefore, the values of all Bernoulli random variables in (5.1) should be determined by the neighbor agent with the shortest trajectory length.

Let $\beta_l^{\min,i+1}(t)=\min\limits_{q,w\in N_l}\left\{\beta_q^{i+1}(t),\beta_w^{i+1}(t)\right\}$ denotes the Bernoulli random variable of the neighboring agent with shortest trajectory length, then one has,

$$\begin{aligned}\hat{\mathbf{c}}_l^{i+1}(t) = (1-\beta_l^{\min,i+1}(t))&\left(\sum_{q\in N_l}\left(\delta\frac{1}{n_q^{i+1}}\sum_{\tau=0}^{i} a_{l,q}^{\tau}(t)\right)\hat{\mathbf{c}}_q^{i}(t)\right.\\ &\left.+\sum_{w\in N_l}\left(\delta\frac{1}{n_w^{i+1}}\sum_{\tau=0}^{i} a_{l,w}^{\tau}(t)\right)\mathbf{c}_w^{i}(t)\right)\\ +\beta_l^{\min,i+1}(t)&\left(\sum_{q\in N_l}a_{l,q}^{i+1}(t)\hat{\mathbf{c}}_q^{i}(t)+\sum_{w\in N_l}a_{l,w}^{i+1}(t)\mathbf{c}_w^{i}(t)\right).\end{aligned} \tag{5.2}$$

**Algorithm 5.1**. Estimate the real coordinates of the follower agent under varying trajectory length

**for** $i = 1$ : $Iteration$.
**initialization** Give the initial iteration coordinate estimation value $\hat{\mathbf{c}}^i(t)$ of the follower agent randomly.
**Step 1.** Each agent transmits its own estimation value $\hat{\mathbf{c}}^i(t)$ to its neighbors.
**Step 2.** Measure the relative distance between agents, and calculate the barycentric coordinate $a^i(t)$.
**Step 3.** Compare $a^i(t)$ with the standard value.
If $\frac{|a^i(t)-a^\kappa(t)|}{a^\kappa(t)} \le \varphi$, then skip to **Step 6**.
**Otherwise,** skip to **Step 4**
**Step 4.** Search out all useful barycentric coordinates from previous iterations $\Lambda_q^{i+1} = \left\{\Gamma_q^j \middle| t \le \Gamma_q^j, j = 1, \ldots, i\right\}$
and $\Lambda_w^{i+1} = \left\{\Gamma_w^j \middle| t \le \Gamma_w^j, j = 1, \ldots, i\right\}$.
**Step 5.** Update
$$\begin{aligned}\hat{\mathbf{c}}_l^{i+1}(t) &= \sum_{q\in N_l}\left(\delta\frac{1}{n_q^{i+1}}\sum_{\tau=0}^{i} a_{l,q}^{\tau}(t)\right)\hat{\mathbf{c}}_q^i(t)\\ &+\sum_{w\in N_l}\left(\delta\frac{1}{n_w^{i+1}}\sum_{\tau=0}^{i} a_{l,w}^{\tau}(t)\right)\mathbf{c}_w^i(t).\end{aligned}$$
**Step 6.** Update
$$\hat{\mathbf{c}}_l^{i+1}(t) = \sum_{q\in N_l} a_{l,q}^{i+1}(t)\hat{\mathbf{c}}_q^i(t) + \sum_{w\in N_l} a_{l,w}^{i+1}(t)\mathbf{c}_w^i(t).$$
**Step 7.** Update $i = i + 1$ and return to **Step 1**.
**end**

The complete process of full historical average data compensation based distributed localization estimation algorithm is given in Algorithm 5.1.

**Remark 5.2** For dynamic multi-agent systems with repetitive operation characteristics, it is natural that the trajectory of the agent is not strictly repeated among different iterations. There will be a certain deviation in the barycentric coordinates calculated at the same sampling time instant in different iteration. Therefore, a constant coefficient $\delta$ close to 1 is added to (5.2). As long as the trajectory of the agent fluctuates within a small reasonable range, the estimation error can converge to a acceptable bounded range by adjusting the value of $\delta$.

### 5.3.4 Convergence Analysis

Consider all followers, one has a compact form,

$$\begin{aligned}\hat{\mathbf{C}}_s^{i+1}(t) &= \left(\left(\mathbf{I}-\boldsymbol{\beta}^{\min,i+1}(t)\right)\mathbf{F}^{i+1}(t)+\boldsymbol{\beta}^{\min,i+1}(t)\mathbf{P}^{i+1}(t)\right)\hat{\mathbf{C}}_s^i(t)\\ &+\left(\left(\mathbf{I}-\boldsymbol{\beta}^{\min,i+1}(t)\right)\mathbf{G}^{i+1}(t)+\boldsymbol{\beta}^{\min,i+1}(t)\mathbf{B}^{i+1}(t)\right)\mathbf{C}_a^i(t),\end{aligned} \tag{5.3}$$

where $\mathbf{C}_a^i(t) = \left[\left(\mathbf{c}_1^i(t)\right)^{\mathrm{T}}, \ldots, \left(\mathbf{c}_m^i(t)\right)^{\mathrm{T}}\right]^{\mathrm{T}}$ are the real coordinate values of all leaders, and $\hat{\mathbf{C}}_s^i(t) = \left[\left(\hat{\mathbf{c}}_{m+1}^i(t)\right)^{\mathrm{T}}, \ldots, \left(\hat{\mathbf{c}}_N^i(t)\right)^{\mathrm{T}}\right]^{\mathrm{T}}$ are the estimated coordinate values of all followers. $\boldsymbol{\beta}^{\min,i+1}(t) = \begin{bmatrix} \beta_{\mathrm{m+1}}^{\min,i+1}(t) & 0 & \cdots & 0 \\ 0 & \beta_{\mathrm{m+2}}^{\min,i+1}(t) & \cdots & 0 \\ \vdots & \vdots & \ddots & \vdots \\ 0 & 0 & \cdots & \beta_N^{\min,i+1}(t) \end{bmatrix}$,

$$\mathbf{F}^{i+1}(t) = \begin{bmatrix} 0 & \frac{\delta}{n_{m+2}^{i+1}} \sum_{\tau=0}^{i} a_{m+1,m+2}^{\tau}(t) & \cdots & \frac{\delta}{n_N^{i+1}} \sum_{\tau=0}^{i} a_{m+1,N}^{\tau}(t) \\ \frac{\delta}{n_{m+1}^{i+1}} \sum_{\tau=0}^{i} a_{m+2,m+1}^{\tau}(t) & 0 & \cdots & \frac{\delta}{n_N^{i+1}} \sum_{\tau=0}^{i} a_{m+2,N}^{\tau}(t) \\ \vdots & \vdots & \ddots & \vdots \\ \frac{\delta}{n_{m+1}^{i+1}} \sum_{\tau=0}^{i} a_{N,m+1}^{\tau}(t) & \frac{\delta}{n_{m+2}^{i+1}} \sum_{\tau=0}^{i} a_{N,m+2}^{\tau}(t) & \cdots & 0 \end{bmatrix}$$

represents the information transfer relationship between followers after taking compensation measures.

$$\mathbf{G}^{i+1}(t) = \begin{bmatrix} \frac{\delta}{n_1^{i+1}} \sum_{\tau=0}^{i} a_{m+1,1}^{\tau}(t) & \frac{\delta}{n_2^{i+1}} \sum_{\tau=0}^{i} a_{m+1,2}^{\tau}(t) & \cdots & \frac{\delta}{n_m^{i+1}} \sum_{\tau=0}^{i} a_{m+1,m}^{\tau}(t) \\ \frac{\delta}{n_1^{i+1}} \sum_{\tau=0}^{i} a_{m+2,1}^{\tau}(t) & \frac{\delta}{n_2^{i+1}} \sum_{\tau=0}^{i} a_{m+2,2}^{\tau}(t) & \cdots & \frac{\delta}{n_m^{i+1}} \sum_{\tau=0}^{i} a_{m+2,m}^{\tau}(t) \\ \vdots & \vdots & \ddots & \vdots \\ \frac{\delta}{n_1^{i+1}} \sum_{\tau=0}^{i} a_{N,1}^{\tau}(t) & \frac{\delta}{n_2^{i+1}} \sum_{\tau=0}^{i} a_{N,2}^{\tau}(t) & \cdots & \frac{\delta}{n_m^{i+1}} \sum_{\tau=0}^{i} a_{N,m}^{\tau}(t) \end{bmatrix}$$

represents the information transfer relationship between followers and leaders after taking compensation measures. $\mathbf{P}^{i+1}(t) = \begin{bmatrix} 0 & a_{m+1,m+2}^{i+1}(t) & \cdots & a_{m+1,N}^{i+1}(t) \\ a_{m+2,m+1}^{i+1}(t) & 0 & \cdots & a_{m+2,N}^{i+1}(t) \\ \vdots & \vdots & \ddots & \vdots \\ a_{N,m+1}^{i+1}(t) & a_{N,m+2}^{i+1}(t) & \cdots & 0 \end{bmatrix}$ denotes the information transfer relationship between followers when agent is running normally.

$$\mathbf{B}^{i+1}(t) = \begin{bmatrix} a_{m+1,1}^{i+1}(t) & a_{m+1,2}^{i+1}(t) & \cdots & a_{m+1,m}^{i+1}(t) \\ a_{m+2,1}^{i+1}(t) & a_{m+2,2}^{i+1}(t) & \cdots & a_{m+2,m}^{i+1}(t) \\ \vdots & \vdots & \ddots & \vdots \\ a_{N,1}^{i+1}(t) & a_{N,2}^{i+1}(t) & \cdots & a_{N,m}^{i+1}(t) \end{bmatrix}$$

denotes the information transfer relationship between followers and leaders when agent is running normally.

**Theorem 5.1** *Consider dynamic multi-agent systems with repeated running characteristics. Applying Algorithm 5.1, the real-time localization problem under varying trajectory length can be solved,* $\mathbb{E}\left\{\left\|\mathbf{e}_l^{i+1}(t)\right\|_\infty\right\}$ *will converge to a bounded value when* $i$ *goes to infinity.*

***Proof*** Define the estimation error as $\mathbf{E}_s^{i+1}(t) = \hat{\mathbf{C}}_s^{i+1}(t) - \mathbf{C}_s^*(t)$ , $\mathbf{E}_a^{i+1}(t) = \mathbf{C}_a^{i+1}(t) - \mathbf{C}_a^*(t)$. Subtracting $\mathbf{C}_s^*(t)$ from both sides of (5.3), and due to $\mathbf{E}_a^i(t) = \mathbf{0}$, one can obtain that

$$\begin{aligned}\mathbf{E}_s^{i+1}(t) = &\begin{pmatrix}\left(\mathbf{I} - \boldsymbol{\beta}^{\min,i+1}(t)\right)\mathbf{F}^{i+1}(t) \\ +\boldsymbol{\beta}^{\min,i+1}(t)\mathbf{P}^{i+1}(t)\end{pmatrix}\mathbf{E}_s^i(t) \\ &+\begin{pmatrix}\left(\mathbf{I} - \boldsymbol{\beta}^{\min,i+1}(t)\right)\mathbf{F}^{i+1}(t) \\ +\boldsymbol{\beta}^{\min,i+1}(t)\mathbf{P}^{i+1}(t) - \mathbf{I}\end{pmatrix}\mathbf{C}_s^*(t) \\ &+\begin{pmatrix}\left(\mathbf{I} - \boldsymbol{\beta}^{\min,i+1}(t)\right)\mathbf{G}^{i+1}(t) \\ +\boldsymbol{\beta}^{\min,i+1}(t)\mathbf{B}^{i+1}(t)\end{pmatrix}\mathbf{C}_a^*(t).\end{aligned} \tag{5.4}$$

Taking norms and applying the expectation operator $\mathbb{E}\{\}$ on both sides of (5.4), one has

$$\begin{aligned}\mathbb{E}\left\{\left\|\mathbf{E}_s^{i+1}(t)\right\|_\infty\right\} \leq &\left\|\begin{pmatrix}\left(\mathbf{I} - \bar{\boldsymbol{\beta}}^{\min,i+1}(t)\right)\mathbf{F}^{i+1}(t) \\ +\bar{\boldsymbol{\beta}}^{\min,i+1}(t)\mathbf{P}^{i+1}(t)\end{pmatrix}\right\|_\infty \mathbb{E}\left\{\left\|\mathbf{E}_s^i(t)\right\|_\infty\right\} \\ +&\left\|\begin{matrix}\begin{pmatrix}\left(\mathbf{I} - \bar{\boldsymbol{\beta}}^{\min,i+1}(t)\right)\mathbf{F}^{i+1}(t)+ \\ \bar{\boldsymbol{\beta}}^{\min,i+1}\mathbf{P}^{i+1}(t) - \mathbf{I}\end{pmatrix}\mathbf{C}_s^*(t) \\ +\begin{pmatrix}\left(\mathbf{I} - \bar{\boldsymbol{\beta}}^{\min,i+1}(t)\right)\mathbf{G}^{i+1}(t) \\ +\bar{\boldsymbol{\beta}}^{\min,i+1}(t)\mathbf{B}^{i+1}(t)\end{pmatrix}\mathbf{C}_a^*(t)\end{matrix}\right\|_\infty.\end{aligned} \tag{5.5}$$

Each element of the matrix $\left(\mathbf{I} - \bar{\boldsymbol{\beta}}^{\min,i+1}(t)\right)\mathbf{F}^{i+1}(t)+\bar{\boldsymbol{\beta}}^{\min,i+1}(t)\mathbf{P}^{i+1}(t)$ is positive and its infinite norm can be expressed as

$$\begin{aligned}&\left\|\left(\left(\mathbf{I} - \boldsymbol{\beta}^{\min,i+1}(t)\right)\mathbf{F}^{i+1}(t)+\boldsymbol{\beta}^{\min,i+1}(t)\mathbf{P}^{i+1}(t)\right)\right\|_\infty = \\ &\max_{l\in\{m+1,\ldots,N\}}\left(\sum_{q\in N_l}\begin{pmatrix}\frac{\delta\left(1-\bar{\beta}_l^{\min,i+1}(t)\right)}{n_q^{i+1}}\sum_{\tau=0}^{i}a_{l,q}^\tau(t) \\ +\bar{\beta}_l^{\min,i+1}(t)a_{l,q}^{i+1}(t)\end{pmatrix}\right),\end{aligned} \tag{5.6}$$

where $\frac{1}{n_q^{i+1}}\sum_{\tau=0}^{i}a_{l,q}^\tau(t)$ is the historical average value of the barycentric coordinates, $a_{l,q}^{i+1}(t)$ is the barycentric coordinates calculated at the current sampling time. Although the trajectory of the agent has a certain deviation among each operation, as long as the trajectory of the agent fluctuates within a small reasonable range, $\frac{\delta}{n_q^{i+1}}\sum_{\tau=0}^{i}a_{l,q}^\tau(t) \leq a_{l,q}^{i+1}(t)$ can be guaranteed by selecting $\delta$ close to 1.

Noted that $0 < \bar{\beta}_l^{\min,i+1}(t) < 1$, one has

$$\frac{\delta\left(1-\bar{\beta}_l^{\min,i+1}(t)\right)}{n_q^{i+1}}\sum_{\tau=0}^{i} a_{l,q}^{\tau}(t)+\bar{\beta}_l^{\min,i+1}(t)a_{l,q}^{i+1}(t)\leq a_{l,q}^{i+1}(t). \tag{5.7}$$

Due to (5.7) and the element in matrix $\left(\mathbf{I}-\bar{\boldsymbol{\beta}}_l^{\min,i+1}(t)\right)\mathbf{F}^{i+1}(t)+\bar{\boldsymbol{\beta}}_l^{\min,i+1}(t)$ $\mathbf{P}^{i+1}(t)$ denotes the barycentric coordinate between followers, yields

$$\max_{l\in\{m+1,\ldots,N\}}\left(\sum_{q\in N_l}\left(\begin{array}{c}\frac{\delta\left(1-\bar{\beta}_l^{\min,i+1}(t)\right)}{n_q^{i+1}}\sum\limits_{\tau=0}^{i} a_{l,q}^{\tau}(t)\\+\bar{\beta}_l^{\min,i+1}(t)a_{l,q}^{i+1}(t)\end{array}\right)\right)<1. \tag{5.8}$$

Then, one has

$$0<\left\|\left(\begin{array}{c}\left(\mathbf{I}-\boldsymbol{\beta}_l^{\min,i+1}(t)\right)\mathbf{F}^{i+1}(t)\\+\boldsymbol{\beta}_l^{\min,i+1}(t)\mathbf{P}^{i+1}(t)\end{array}\right)\right\|_{\infty}=\sigma_1<1. \tag{5.9}$$

Similarly, due to $\mathbf{c}_l^*(t), l\in\{1,\ldots,N\}$ represents the real coordinate of the agent, $\mathbf{c}_l(t)=a_{l,h}(t)\mathbf{c}_h(t)+a_{l,j}(t)\mathbf{c}_j(t)+a_{l,k}(t)\mathbf{c}_k(t)$ and $\frac{\delta\left(1-\bar{\beta}_l^{\min,i+1}(t)\right)}{n_q^{i+1}}\sum\limits_{\tau=0}^{i} a_{l,q}^{\tau}(t)+$ $\bar{\beta}_l^{\min,i+1}(t)a_{l,q}^{i+1}(t)\leq a_{l,q}^{i+1}(t)$, it follows from the last two items on the right side of (5.4) that

$$\begin{array}{c}\left(\begin{array}{c}\frac{\left(1-\bar{\beta}_{m+1}^{\min,i+1}(t)\right)\delta}{n_1^{i+1}}\sum\limits_{\tau=0}^{i} a_{m+1,1}^{\tau}(t)\\+\bar{\beta}_{m+1}^{\min,i+1}(t)a_{m+1,1}^{i+1}(t)\end{array}\right)\mathbf{c}_1^*(t)+\cdots+\\
\left(\begin{array}{c}\frac{\delta\left(1-\bar{\beta}_{m+1}^{\min,i+1}(t)\right)}{n_N^{i+1}}\sum\limits_{\tau=0}^{i} a_{m+1,N}^{\tau}(t)\\+\bar{\beta}_{m+1}^{\min,i+1}(t)a_{m+1,N}^{i+1}(t)\end{array}\right)\mathbf{c}_N^*(t)\leq\mathbf{c}_{m+1}^*(t),\\
\vdots\\
\left(\begin{array}{c}\frac{\left(1-\bar{\beta}_{N}^{\min,i+1}(t)\right)\delta}{n_1^{i+1}}\sum\limits_{\tau=0}^{i} a_{N,1}^{\tau}(t)+\\\bar{\beta}_{N}^{\min,i+1}(t)a_{N,1}^{i+1}(t)\end{array}\right)\mathbf{c}_1^*(t)+\cdots+\\
\left(\begin{array}{c}\frac{\delta\left(1-\bar{\beta}_{N}^{\min,i+1}(t)\right)}{n_{N-1}^{i+1}}\sum\limits_{\tau=0}^{i} a_{N,N-1}^{\tau}(t)\\+\bar{\beta}_{N}^{\min,i+1}(t)a_{N,N-1}^{i+1}(t)\end{array}\right)\mathbf{c}_{N-1}^*(t)\leq\mathbf{c}_N^*(t).\end{array} \tag{5.10}$$

Which yields,

$$\left\| \begin{array}{l} \begin{pmatrix} \left(\mathbf{I}-\bar{\boldsymbol{\beta}}^{\min,i+1}(t)\right)\mathbf{F}^{i+1}(t) \\ +\bar{\boldsymbol{\beta}}^{\min,i+1}\mathbf{P}^{i+1}(t)-\mathbf{I} \end{pmatrix} \mathbf{C}_s^*(t) \\ +\begin{pmatrix} \left(\mathbf{I}-\bar{\boldsymbol{\beta}}^{\min,i+1}(t)\right)\mathbf{G}^{i+1}(t) \\ +\bar{\boldsymbol{\beta}}^{\min,i+1}(t)\mathbf{B}^{i+1}(t) \end{pmatrix} \mathbf{C}_a^*(t) \end{array} \right\|_\infty \leq \sigma_2, \tag{5.11}$$

where $\sigma_2 > 0$ is a bounded constant, and the value of $\sigma_2$ depend on the magnitude of the trajectory deviation. As long as the trajectory deviation within a small reasonable range, $\sigma_2$ is an acceptable value. The range of allowable trajectory deviation varies with different application scenarios.

Based on (5.6)–(5.11), (5.5) can be expressed as,

$$\begin{array}{l} \mathbb{E}\left\{\left\|\mathbf{E}_s^{i+1}(t)\right\|_\infty\right\} \leq \sigma_1\mathbb{E}\left\{\left\|\mathbf{E}_s^{i}(t)\right\|_\infty\right\}+\sigma_2 \\ \leq \cdots \leq \sigma_1^{i+1}\mathbb{E}\left\{\left\|\mathbf{E}_s^{0}(t)\right\|_\infty\right\}+\frac{\sigma_2}{1-\sigma_1} \end{array}. \tag{5.12}$$

Due to the fact that $0 < \sigma_1 < 1$, $\sigma_2 > 0$ is a bounded constant and the initial estimation error $\left\|\mathbf{E}_s^0(t)\right\|_\infty$ is bounded, (5.12) implies that when $i$ goes to infinity, $\mathbb{E}\left\{\left\|\mathbf{e}_l^{i+1}(t)\right\|_\infty\right\}$ will converge to a small bounded value $\frac{\sigma_2}{1-\sigma_1}$.

The proof of Theorem 5.1 is complete.

**Remark 5.3** If the trajectory of the agent is completely repeated in each operation, the historical data can compensate the missing trajectory data without error. Then the proposed compensation algorithm can guarantee that the estimation error converge to zero under varying trajectory length. Even if the trajectory of the agent is not completely repeated and has a small deviation, the proposed compensation algorithm also can guarantee the coordinate estimation error converges to a small bounded value.

## 5.4 Extensions to Receding Horizon Historical Average Data Compensation

The advantage of Algorithm 5.1 with full historical average data compensation is that the historical database is constantly supplemented with the increase of iterations. It can weaken the influence of non-repeated trajectory on estimation accuracy. However, with the increase of iterations, the computation burden of the system increases gradually, then the convergence rate of the estimation error will slow down.

### 5.4.1 Receding Horizon Historical Average Data Compensation

In this section, a receding horizon historical average data compensation based distributed estimation algorithm is designed to reduce the computation burden.

Define $\Lambda_q^{i+1} = \left\{\Gamma_q^j \middle| t \le \Gamma_q^j, j = i - \nu + 1, \ldots, i\right\}$, $\Lambda_w^{i+1} = \left\{\Gamma_w^j \middle| t \le \Gamma_w^j, j = i - \nu + 1, \ldots, i\right\}$. $\upsilon > 1$ is a positive constant, it represents the size of receding horizon. $n_q^{i+1}$ and $n_w^{i+1}$ represent the number of elements in $\Lambda_q^{i+1}$ and $\Lambda_w^{i+1}$, respectively.

Before design the compensation algorithm, an assumption is given as follows.

**Assumption 5.2** For the sampling time instant $t \in \{0, 1, \ldots, \Gamma^i\}$ in the $i$th iteration, there exists at least one iteration whose length $\Gamma^j$, $j \in \{i - \upsilon, \ldots, i - 1\}$ is greater than $t$ in previous iterations.

A receding horizon historical data compensation based distributed localization estimation algorithm is proposed.

$$\begin{aligned}\hat{\mathbf{c}}_l^{i+1}(t) = (1 - \beta_l^{\min,i+1}(t))&\left(\sum_{q\in N_l}\left(\delta\frac{1}{n_q^{i+1}}\sum_{\tau=i-\upsilon+1}^{i} a_{l,q}^{\tau}(t)\right)\right.\\ &\left.\times\hat{\mathbf{c}}_q^i(t) + \sum_{w\in N_l}\left(\delta\frac{1}{n_w^{i+1}}\sum_{\tau=i-\upsilon+1}^{i} a_{l,w}^{\tau}(t)\right)\mathbf{c}_w^i(t)\right)\\ +\beta_l^{\min,i+1}(t)&\left(\sum_{q\in N_l} a_{l,q}^{i+1}(t)\hat{\mathbf{c}}_q^i(t) + \sum_{w\in N_l} a_{l,w}^{i+1}(t)\mathbf{c}_w^i(t)\right).\end{aligned} \tag{5.13}$$

**Remark 5.4** The difference between (5.13) and (5.2) is that all historical information is used to compensate in (5.2), while only the recent historical data is used in (5.13), which reduces the computation burden.

The process of localization estimation Algorithm 5.2 is shown below,

**Step 1–3** and **Step 6–7** are the same with Algorithm 5.1.

**Step 4.** Search out all useful barycentric coordinates from the recent iterations $\Lambda_q^{i+1} = \left\{\Gamma_q^j \middle| t \le \Gamma_q^j, j = i - \nu + 1, \ldots, i\right\}$ and $\Lambda_w^{i+1} = \left\{\Gamma_w^j \middle| t \le \Gamma_w^j, j = i - \nu + 1, \ldots, i\right\}$.

**Step 5.** Update
$$\begin{aligned}\hat{\mathbf{c}}_l^{i+1}(t) = &\sum_{q\in N_l}\left(\delta\frac{1}{n_q^{i+1}}\sum_{\tau=i-\upsilon+1}^{i} a_{l,q}^{\tau}(t)\right)\hat{\mathbf{c}}_q^i(t)\\ &+\sum_{w\in N_l}\left(\delta\frac{1}{n_w^{i+1}}\sum_{\tau=i-\upsilon+1}^{i} a_{l,w}^{\tau}(t)\right)\mathbf{c}_w^i(t).\end{aligned}$$

We omit the complete process of Algorithm 5.2 here.

### 5.4.2 Convergence Analysis

**Theorem 5.2** *Consider dynamic multi-agent systems with repeated running characteristics. Applying Algorithm 5.2, the real-time localization problem under varying trajectory length can be solved,* $\mathbb{E}\left\{\left\|\mathbf{e}_l^{i+1}(t)\right\|_\infty\right\}$ *will converge to a bounded value when i goes to infinity.*

***Proof*** The proof process is similar to the Theorem 5.1, which is omitted here.

### 5.4.3 Corollary

The integration of distributed localization and formation control of multi-agent systems is considered in this subsection.

Consider a system with $N$ nonholonomic mobile vehicles. The kinematic model of each vehicle can be formulated as

$$\begin{aligned} \dot{c}_{l,1} &= v_l \cos\theta_l, \\ \dot{c}_{l,2} &= v_l \sin\theta_l, \\ \dot{\theta}_l &= \omega_l, \end{aligned} \tag{5.14}$$

where $l = 1, \ldots, N$, $\mathbf{c}_l = \left[c_{l,1}, c_{l,2}\right]^{\mathrm{T}}$ and $\theta_l$ represent the position coordinate and the orientation angle of the agent $l$ in the global frame, respectively. $v_l$ is the linear velocity and $\omega_l$ denotes the angular velocity.

The relative measurement is shown in Fig. 5.2. {Re , Im} denotes the global frame and {Re_$l$, Im_$l$} is the local frame of the agent $l$. $\theta_{lz} = \theta_l - \theta_z$ represents the relative orientation angle of the agent $l$ in the local frame of the agent $z$.

Define $\theta_{lz}^* \triangleq \theta_l^* - \theta_z^*$ denotes the desired relative angular spacing between the agent $l$ and the agent $z$, where $\left|\theta_l^*\right| \in [0, 2\pi]$ and $\left|\theta_z^*\right| \in [0, 2\pi]$. The details are shown in Fig. 5.3, where $r$ is radius of the circle.

In this subsection, the circular formation problem is considered. The objective is to design an integrated scheme of localization and formation control, all agents cooperatively estimate the positions and achieve a prescribed circular formation by using the obtained position estimates.

$$\begin{aligned} \lim_{i\to\infty} \mathbf{c}_l^i(t) - \mathbf{c}_p &= r\left[\sin\theta_l^i(t), -\cos\theta_l^i(t)\right]^{\mathrm{T}}, \\ \lim_{i\to\infty} \omega_l^i(t) &= \omega^*(t), \\ \lim_{i\to\infty} \theta_{lz}^i(t) &= \theta_{lz}^*(t), \end{aligned} \tag{5.15}$$

where $i$ represents the iteration number, $\mathbf{c}_p = \left[c_{p,1}, c_{p,2}\right]^{\mathrm{T}}$ is the given center of the circle, $\omega^*$ represents the desired angular velocity, and the desired position $\mathbf{c}_l^*(t) = \left[c_{l,1}^*(t), c_{l,2}^*(t)\right]^{\mathrm{T}}$ can be calculated by (5.15).

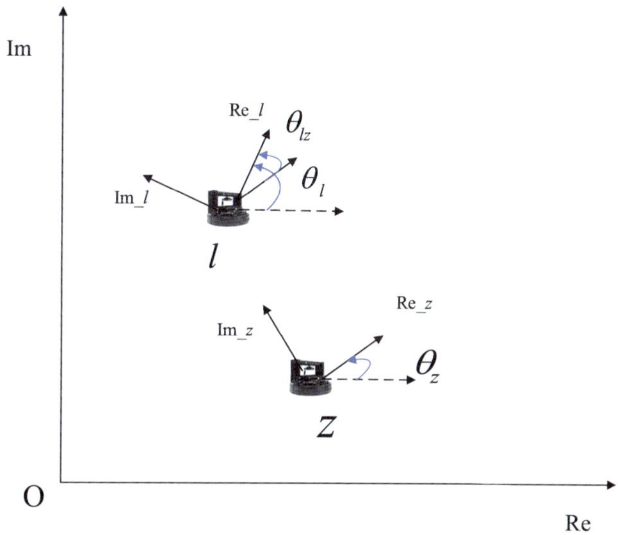


**Fig. 5.2** The relative measurement in local frame

**Fig. 5.3** Desired angular spacing in circular formation

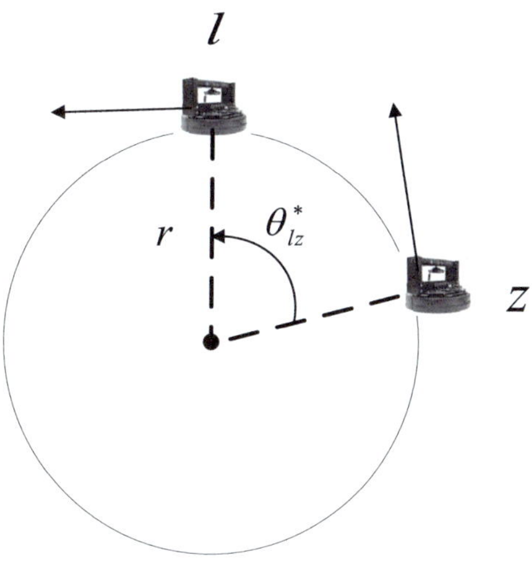


For controller design, an assumption is given below.

**Assumption 5.3** The given center $\mathbf{c}_p$ of the circular can be measured by the agents in their local frames. The desired angular velocity $\omega^*$ and the radius $r$ of the circle, are known to all agents.

An integrated scheme of localization and formation control based on the iterative learning is proposed as follows.

$$
\begin{aligned}
\omega_l^i(t) &= K_1 \sum_{z=1}^{N} \lambda_{lz} \left(\theta_{lz}^i(t) - \theta_{lz}^*(t)\right) + \omega^*(t) \\
v_l^i(t) &= \omega_l^i(t) r + \omega_l^i(t) \hat{e}_{l,2}^i(t) + K_2 \hat{e}_{l,1}^i(t) - K_3 \hat{e}_{l,2}^i(t) \omega_l^i(t)
\end{aligned}
\tag{5.16}
$$

where $\hat{e}_{l,1}^i(t) = c_{l,1}^*(t) - \hat{c}_{l,1}^i(t)$ and $\hat{e}_{l,2}^i(t) = c_{l,2}^*(t) - \hat{c}_{l,2}^i(t)$ are the position estimation error on the real axis and the imaginary axis, respectively. $\lambda_{lz} > 0$ represents the communication weight between agents $l$ and $z$. $K_\tau > 0$, $\tau = 1, 2, 3$ are positive constants.

## 5.5 Simulation and Experiment

### 5.5.1 Numerical Simulation

The dynamic multi-agent systems with seven agents is considered to test the validity of the proposed distributed localization algorithm with compensation. The directed topology is shown in Fig. 5.4, and the trajectory curves of the agents are shown in Fig. 5.5.

This section will discuss the convergence results of multi-agent systems with randomly varying trajectory lengths in three different cases. The first case is that the historical average data compensation under the condition of incompletely repeated trajectory. The second case is that the historical average data compensation under the condition of completely repeated trajectory. The third case is that there is no historical average data compensation under the condition of incompletely repeated trajectory.

The random disturbance is used to simulate the incomplete repetition characteristic of the agent trajectory. The Bernoulli random variables is used to represent the

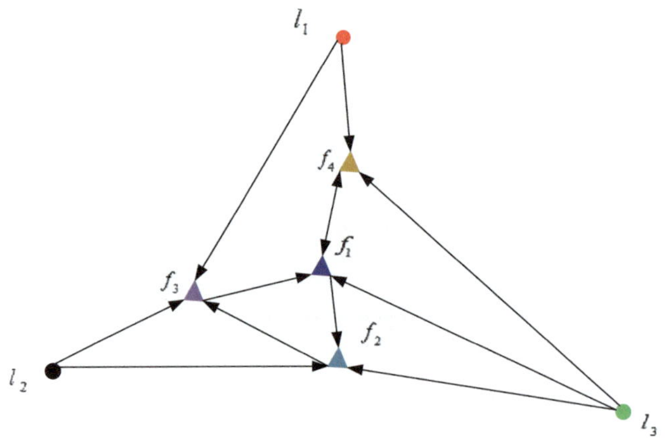


**Fig. 5.4** The directed topology of the multi-agent systems

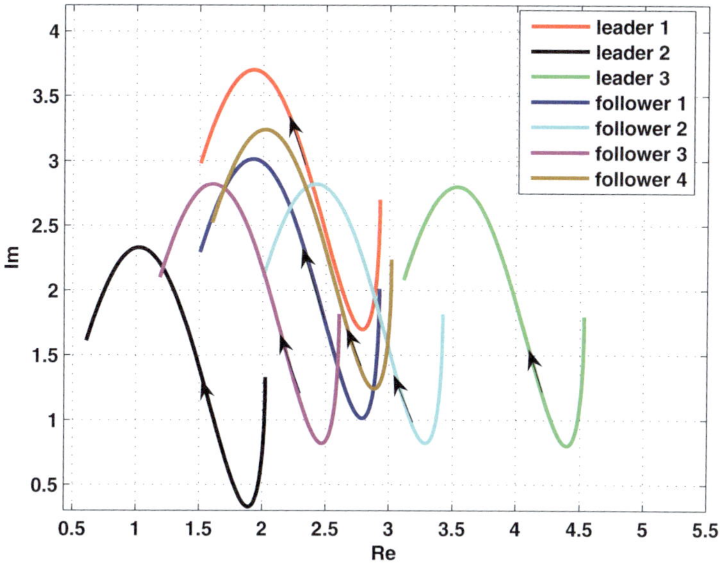


**Fig. 5.5** The real trajectories of the multi-agent systems

random variation of the trajectory length. The initial estimated values of the trajectory curve of all follower agents are given randomly.

**Case 1.** Historical average data compensation under the condition of incompletely repeated trajectory.

In this case, we applied random disturbance $0.1 * \text{rand}$ on the virtual axis and the real axis of the agent trajectory. The probabilities of varying trajectory length of follower agents 1, 2, 3 and 4 are set as 10%, 15%, 15% and 10% respectively. Select $\delta = 0.95$ and $\upsilon = 3$, then, applying the proposed Algorithms 5.1 and 5.2.

The simulation results are shown in Figs. 5.6, 5.7, 5.8 and 5.9. Figure 5.6 shows the estimation process of all follower agents under Algorithm 5.1. One can find that although the initial trajectory curve is far from the real trajectory, the estimator can estimate the approximate exact trajectory curve after 15 iterations under Algorithm 5.1. Figures 5.7 and 5.8 shows the maximum estimation error along the iteration axis under Algorithms 5.1 and 5.2, respectively. The maximum estimation error is defined as $e_{l,\max}(i) = \max_t \left\{ \left\| \mathbf{c}_l^i(t) - \mathbf{c}_i^*(t) \right\|_2 \right\}$, where $t$ represents all sampling times of the $i$th iteration. It can be concluded that the proposed Algorithms 5.1 and 5.2 can guarantee the asymptotic convergence of the estimation error and achieve the approximate exact real-time localization under the condition of incompletely repeated trajectory. Figure 5.9 shows the comparison results of the maximum estimation error of four follower agents after estimator converges. The bright blue histogram represents the maximum estimation error under Algorithm 5.1, the red histogram represents the maximum estimation error under Algorithm 5.2. As can be seen from Fig.

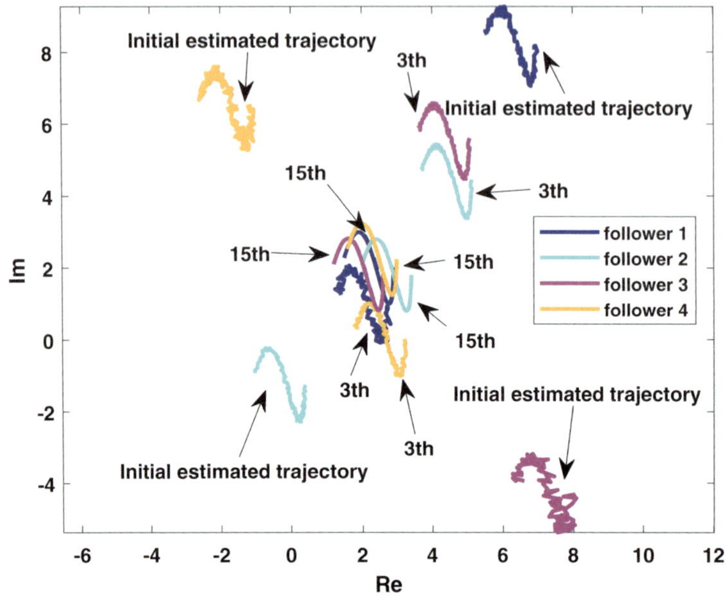


Fig. 5.6 The trajectory estimation process of four follower agents under Algorithm 5.1

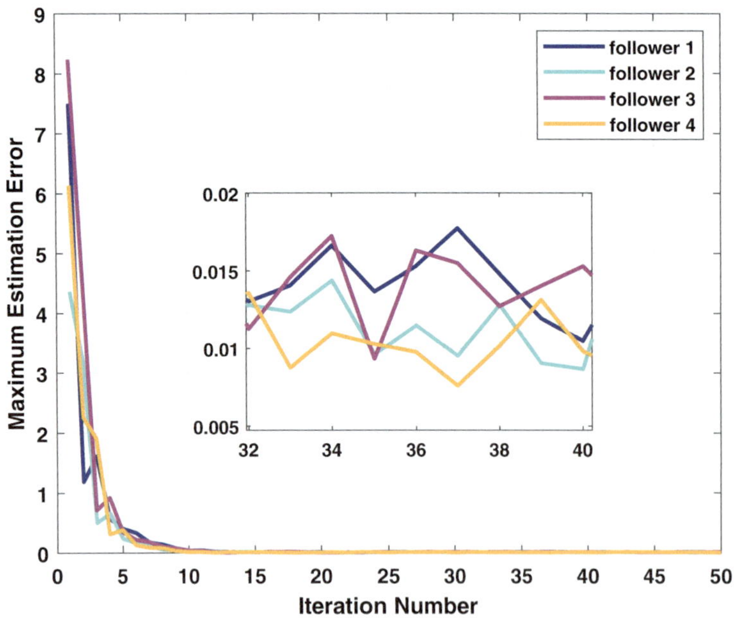


Fig. 5.7 The maximum estimation error of four follower agents under Algorithm 5.1

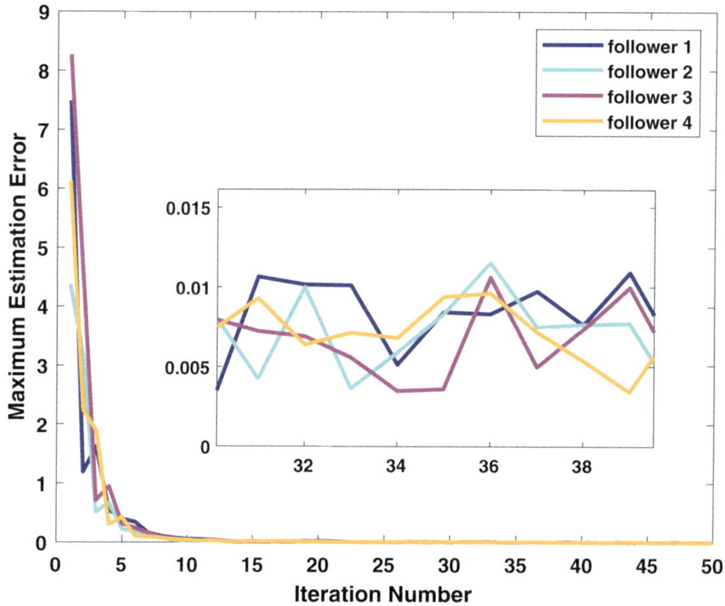


**Fig. 5.8** The maximum estimation error of four follower agents under Algorithm 5.2

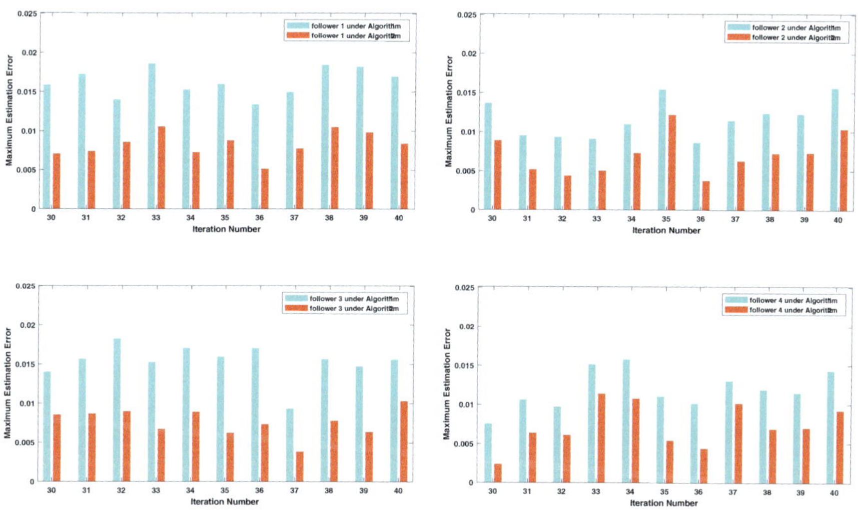


**Fig. 5.9** The comparison results of follower agents after estimator convergence under Algorithm 5.1 and Algorithm 5.2

5.9, the maximum estimation error under Algorithm 5.1 is larger than that under Algorithm 5.2.

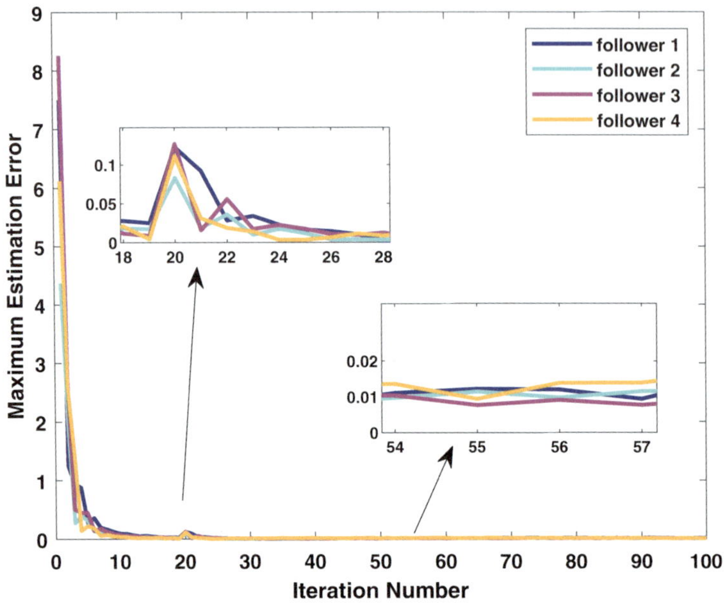


**Fig. 5.10** The maximum estimation error of four follower agents under Algorithm 5.1 when encountering unexpected disturbance

In order to distinguish the convergence effect of Algorithm 5.1 and Algorithm 5.2, some new simulation conditions are set. After the estimation error converges, in the 20th iteration, a random disturbance $0.2 * \text{rand}$ is imposed on the real axis and virtual axis of the agent trajectory, and other conditions remain unchanged. The simulation results are shown in Figs. 5.10 and 5.11.

Compare Figs. 5.10 and 5.11, it can be found that the maximum estimation error is not fluctuating greatly when Algorithm 5.1 is applied. But the maximum estimation error has a large fluctuation when Algorithm 5.2 is applied, and the fluctuation will also have an effect on the next few iterations.

From Figs. 5.6, 5.7, 5.8, 5.9, 5.10 and 5.11, one can conclude that the proposed Algorithm 5.1 with full historical average data compensation can weaken the effect of unexpected random disturbance on the estimation error, but its final maximum estimation error will be affected by the random disturbances from previous iterations. It is suitable for multi-agent systems that require low localization accuracy and strong anti-interference ability. The proposed Algorithm 5.2 with receding horizon historical average data compensation is more sensitive to unexpected random disturbance. Yet its final localization accuracy is higher than that of Algorithm 5.1, and the computation burden is much less than that of Algorithm 5.1. It is suitable for multi-agent systems that require high localization accuracy and low external interference.

**Case 2.** Historical average data compensation under the condition of completely repeated trajectory.

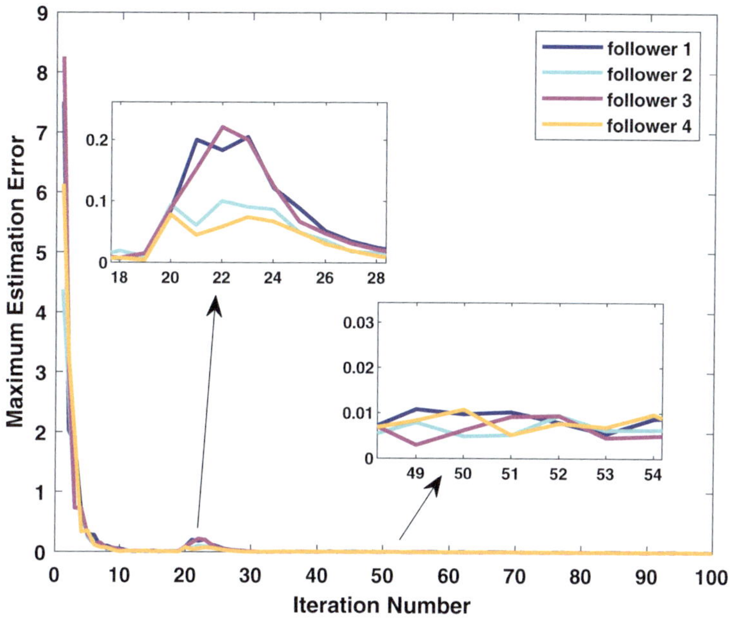


**Fig. 5.11** The maximum estimation error of four follower agents under Algorithm 5.2 when encountering unexpected disturbance

In this case, the trajectory of the agent is completely repeated. The probabilities of trajectory length change of follower agents 1, 2, 3 and 4 are the same with Case 1. Select $\delta = 1$, then, applying the Algorithm 5.1. The estimation error curves under Algorithm 5.2 are omitted to save space.

Figure 5.12 shows the maximum estimation error curves in the estimation process along the iteration axis. One can find that the estimation error will converge to zero when the trajectory is completely repeated.

From the simulation results of the Cases 1 and 2, the following conclusions are summarized. First, the two distributed localization estimation algorithms based on historical average data compensation can guarantee that the estimation error converges to zero for multi-agent systems with completely repeated operation characteristics. Second, as long as the trajectory deviation fluctuates in a small reasonable range, the estimation algorithms can still guarantee the asymptotic convergence of the estimation error and the estimation error can converge to a small bounded range, even if the trajectories of multi-agent systems is not completely repeated. Third, Algorithm 5.1 based on full historical average data compensation can weaken the effect of unexpected random disturbance on the estimation error, but with the increase of the iterations, the amount of calculation will become larger. Algorithm 5.2 based on receding horizon historical average data compensation can reduce computation burden effectively, but it is particularly sensitive to the unexpected random disturbance.

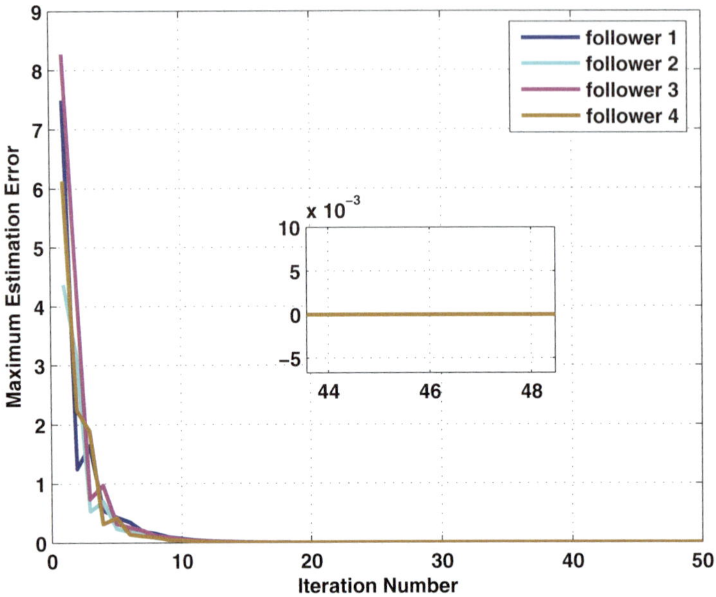


**Fig. 5.12** The maximum estimation error of the follower agents under completely repeated trajectory

### 5.5.2 *QBot-2e Robot Experiment*

Six QBot-2e robots produced by Quanser are used to test the validity of the proposed distributed localization algorithm with compensation. The Qbot-2e robot creates communication interface with Matlab/Simulink through real-time control software QUARC.

The real coordinates of the six robots can be accurately obtained through the self-built map and cameras. The leader robot directly uses the real coordinates, and the follower robot estimates its own coordinates based on the relative distance. Figures 5.13 and 5.14 shows the QBot-2e robot experimental platform and the deployment of six robots. The directed graph and the trajectory curves of six robots are shown in Figs. 5.15 and 5.16. The three robots in outer circle are leaders, and the other three robots in inner circle are followers to be located.

This section will discuss the estimation results of three follower robots under two different cases. The initial estimated values of the trajectory curves are given randomly. The velocity of the robot is 0.5 m/s, the unit is meter and the sampling frequency is 60 Hz. The percentage value $\varphi$ is set to 15%.

**Case 1.** Historical average data compensation.

In the process of experiment, the trajectory of the robot fluctuates greatly in initial running. With the increase of the operation, the trajectory of the robot will gradually become stable. However, affected by the road roughness, the sensor measurement

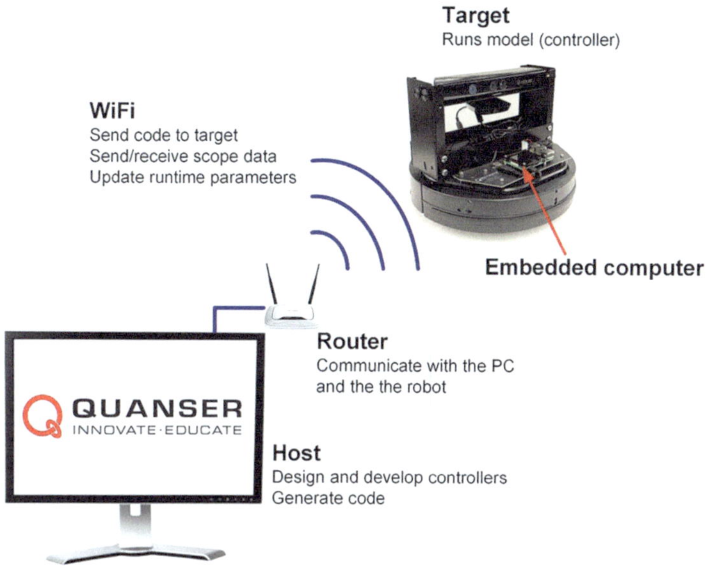


**Fig. 5.13** The QBot-2e robot experimental platform

**Fig. 5.14** The deployment of all robots

accuracy and the robot's own mechanical structure, the trajectory of the robot cannot be guaranteed to be completely repeated each operation.

The real trajectory curve of robot 1 is shown in Fig. 5.17. For brevity, only part of the trajectory data of robot 1 is presented, and the difference of trajectory length between different iterations can be clearly distinguished through the trajectory curve. Select $\delta = 0.95$ and $\upsilon = 5$, then, applying Algorithms 5.1 and 5.2.

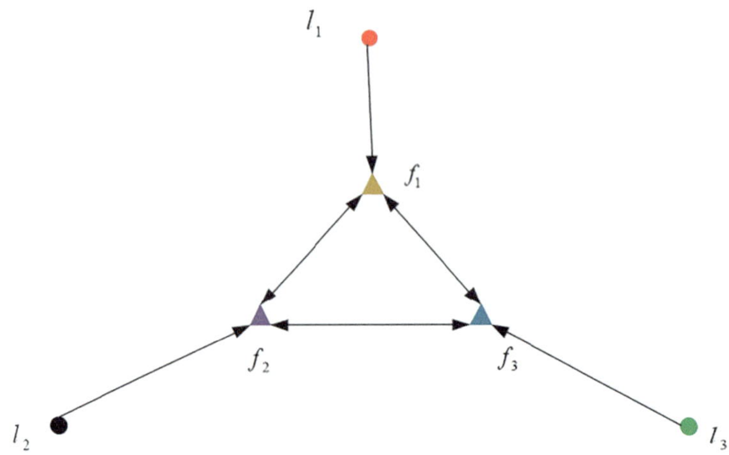


**Fig. 5.15** The directed graph of all robots

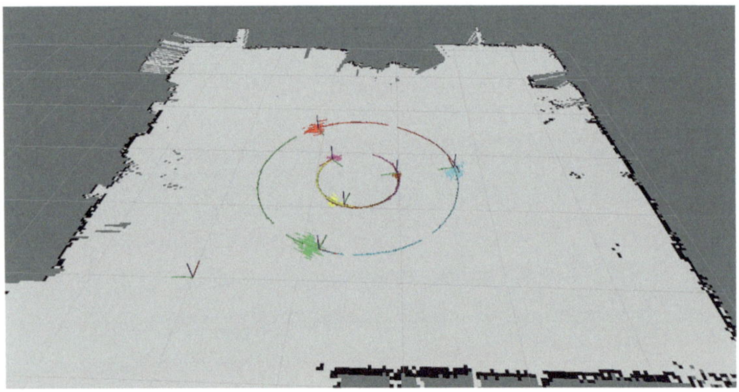

**Fig. 5.16** The real trajectory curves of all robots

Figure 5.18 is the estimation process of three follower robots under Algorithm 5.1. Although the initial estimation curve is given randomly, an approximate accurate estimation results can be guaranteed with the increase of iteration. Figures 5.19 and 5.20 shows the maximum estimation error curves along the iteration axis after applying Algorithms 5.1 and 5.2, respectively. Both these two algorithms can ensure the asymptotic convergence of the estimation error and achieve the approximate exact real-time localization under the incomplete repetitive trajectory.

Figure 5.21 shows the comparison results of the maximum estimation error of three robots after the estimator converges. The bright blue histogram represents the maximum estimation error under Algorithm 5.1, and the red histogram represents the maximum estimation error under Algorithm 5.2. The maximum estimation error under Algorithm 5.1 is larger than that under Algorithm 5.2.

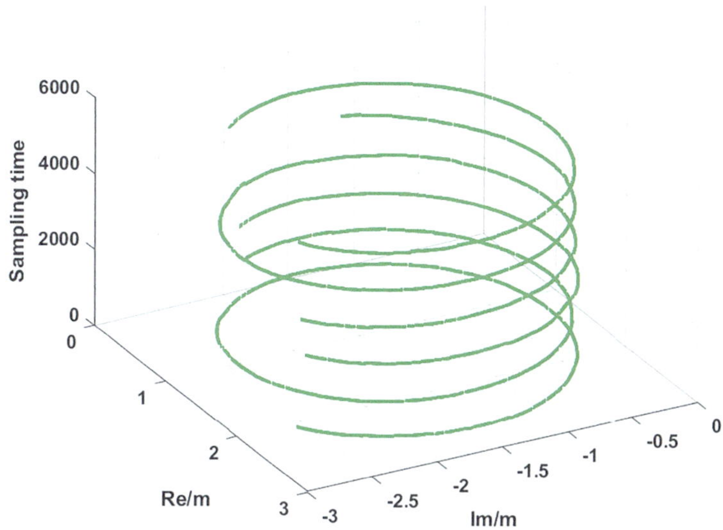


**Fig. 5.17** The real trajectory curve of robot 1 under Algorithm 5.1

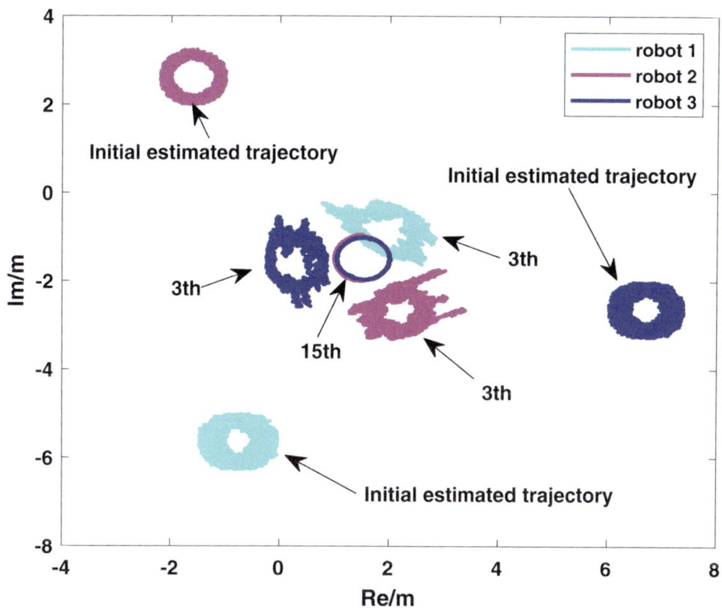


**Fig. 5.18** The trajectory estimation process of three follower robots under Algorithm 5.1

From Figs. 5.19, 5.20 and 5.21, the proposed methods with historical average data compensation can effectively deal with the persistent real-time localization problem

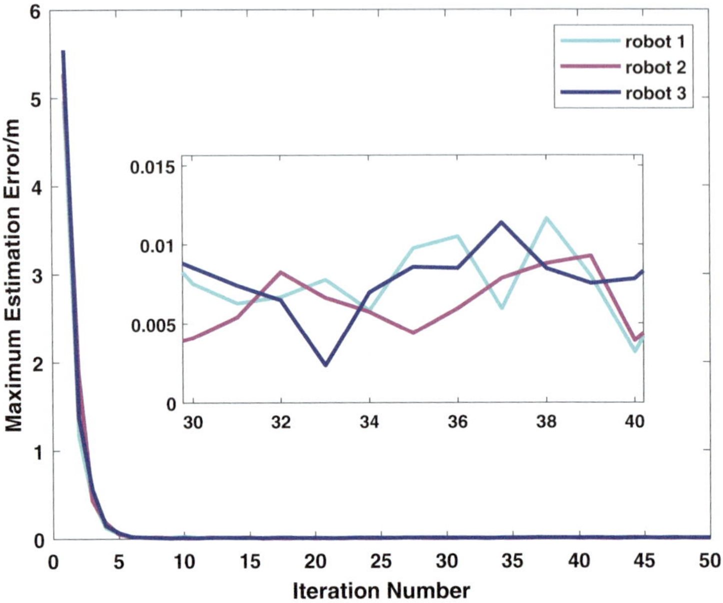


Fig. 5.19 The maximum estimation error of three follower robots under Algorithm 5.1

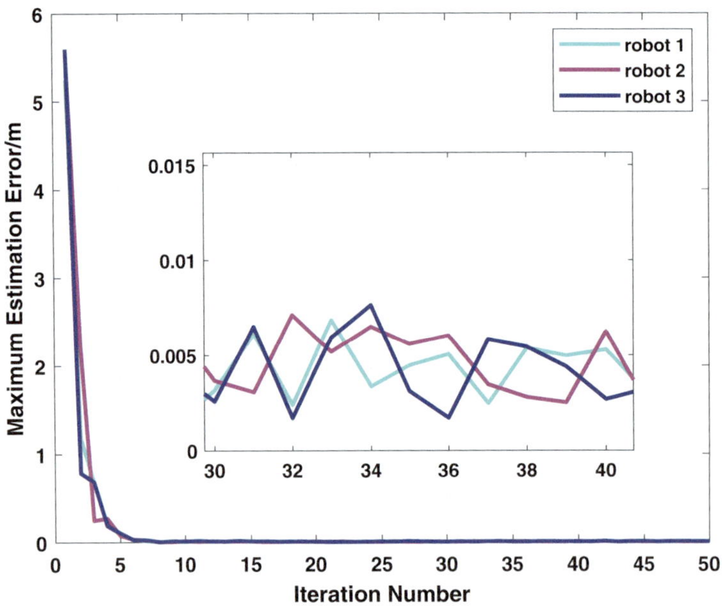


Fig. 5.20 The maximum estimation error of three follower robots under Algorithm 5.2

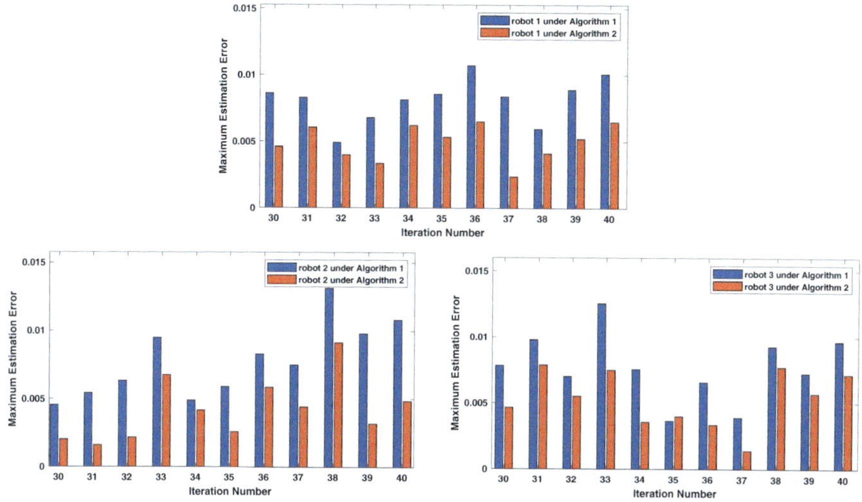


**Fig. 5.21** The comparison results of three follower robots after estimator convergence under Algorithm 5.1 and Algorithm 5.2

of robot with randomly varying trajectory lengths. Even if the trajectory of the robot is not completely repeated, it can guarantee approximate exact localization result.

**Case 2.** Without historical average data compensation.

In this case, the parameters of estimator are same with Case 1. When the trajectory of the robot is incomplete, the estimator does not take any compensation measures. It will stop updating and keep the estimated value of the last iteration unchanged. The maximum estimation error curves are shown in Fig. 5.22. The convergence rate of the estimator is slow down significantly without any compensation measures.

The following conclusions can be drawn from the experimental results. First, the proposed two distributed localization estimation algorithms can effectively alleviate the influence of the randomly trajectory length on the convergence speed of the estimator. Second, even if the trajectory of the robot is not completely repeated, the proposed estimation method can still guarantee approximate accurate localization results.

## 5.6 Conclusion

This chapter solves the persistent real-time localization problem of dynamic multi-agent systems with repetitive running characteristics and randomly varying trajectory lengths under directed graph. Two novel distributed iterative learning localization estimation methods with historical average data compensation is proposed. The designed diagnosis mechanism and search mechanism can meet the requirements of

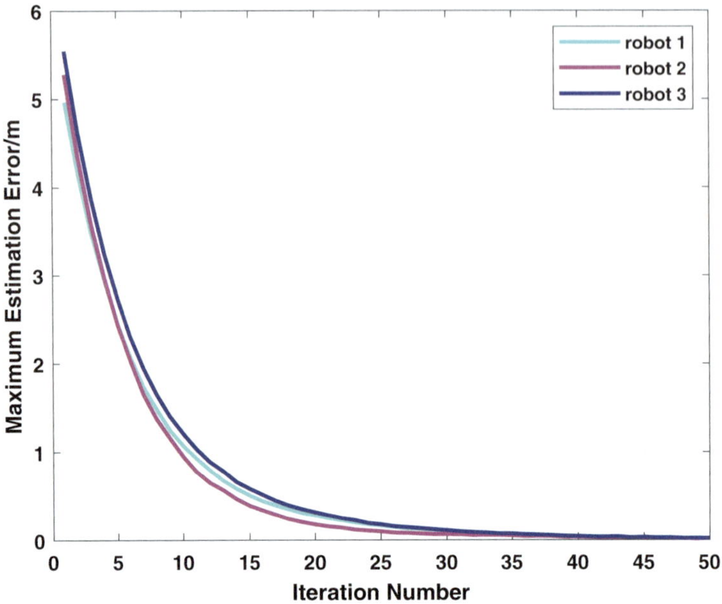


**Fig. 5.22** The maximum estimation error of three follower robots without compensation

practical application. The distributed localization estimation algorithm based on full historical average data compensation can effectively weaken the influence of incomplete repeated trajectories on the estimation error, and make the estimator achieve the purpose of asymptotic convergence. The improved distributed localization estimation algorithm based on receding horizon historical average data compensation can greatly reduce the computation burden. The convergence of the estimation methods in the sense of mathematical expectation is derived through the rigorous analysis. The numerical simulation and the experimental result based on QBot-2e robot platform verify the effectiveness of the proposed methods.

## References

1. Liang, J., Bu, X., Wang, Q., et al.: Iterative consensus control for a class of nonlinear multi-agent systems with randomly varying iteration lengths. In: Proceedings of IEEE International Control Automation, pp. 1014–1019 (2019)
2. Liu, Y., Fan, Y., Jia, Y.: Iterative learning formation control for continuous-time multi-agent systems with randomly varying trial lengths. J. Franklin Inst. **357**(14), 9268–9287 (2020)

# Chapter 6
# Data-Driven Adaptive Distributed Localization of MASs with Sensor Failure

This chapter solves the localization estimation of dynamic multi-agent systems (MASs) with sensor multiplicative failures, which is more general yet challenging to address than static sensor networks with ideal conditions. Barycentric coordinate is introduced to characterize the relative positions between agents. A new linear data model is constructed to represent the relationship between barycentric coordinates and relative distance. Based on the linear model, an adaptive parameter estimation algorithm is designed, and then it is applied to solve the relative distance compensation problem of the MASs with sensor multiplicative failures. Using the estimated parameter, a data-driven adaptive distributed localization estimation scheme based on iterative learning is proposed, in which only the measured relative distance data is available instead of the system model information. Key to obtaining accurate localization is overcoming the difficulties from inaccurate relative distance variables due to sensor failure via the data-driven adaptive relative distance compensation method. The numerical examples verify the effectiveness of the proposed methods.

## 6.1 Introduction

Most existing localization methods assume that the sensor works in good condition and in the ideal application environment. However, sensor failure, as one of the main causes of unstable and poor performance, is more frequent than actuators and often arises due to the complexity of the application environment. The localization estimation performance also depends on the quality of the information measured by the sensors. The control problem of systems with sensor failure has attracted more attention in industrial and academic communities. Some sensor failure detection and compensation algorithms have been proposed [1–3], to name a few. The effect of sensor failure on the convergence and accuracy of localization estimation methods has yet to be fully considered. Localization estimation problem under sensor failure

 
Y. Lv et al., *Distributed Localization and Formation Control of Multi-Agent Systems with Local Sensing Information*, https://doi.org/10.1007/978-981-96-6015-5_6

is usually challenging to deal with mainly because there is no feedback information except the measured variables. In this work, the model of the agent is unknown, which means that the state information of the agent cannot be obtained. For example, the Kalman filter algorithm (which requires both state information and measurement information) cannot be applied to solve the localization estimation problem in the case of sensor failure. Hence, the investigation of measurement variable compensation is of particular importance for accurate localization estimation.

Specifically, a data-driven adaptive parameter estimation algorithm is designed based on the constructed linear data model. Based on iterative learning and combined with the parameter estimation algorithm, a data-driven adaptive distributed localization estimation scheme with a compensation mechanism is proposed, which is used to overcome the localization problem of MASs with sensor multiplicative failures.

The main contribution can be summarized as follows:

(1) The localization estimation problem of dynamic MASs with repetitive operation characteristics is solved compared with the static network localization.
(2) A linear data model is constructed to describe the relationship of barycentric coordinates and relative distance in iteration domain.
(3) The effect of sensor failure on the final localization accuracy is eliminated. A data-driven adaptive distributed localization estimation scheme with compensation mechanism is proposed.

The remaining of this chapter is shown below: Sect. 6.2 gives the problem formulation. Section 6.3 proposes the data-driven adaptive distributed localization scheme. Section 6.4 provides the convergence analysis. The simulation example is given in Sect. 6.5. Some summaries are provided in Sect. 6.6.

## 6.2 Problem Formulation

Consider a MAS containing $N$ agents that can perform tasks repeatedly, where $m$ are leaders and other $(N-m)$ are followers. The leader agents get the global coordinates through the equipped locating device, while the follower agents can estimate their coordinates by measuring the relative distance and communicating directly or indirectly with the leader agents.

A distributed method is designed to solve the localization problem of multi-agent systems under ideal environment.

$$\hat{\mathbf{c}}_l^{i+1}(t) = \sum_{q \in \mathcal{N}_l} a_{l,q}^{i+1}(t)\hat{\mathbf{c}}_q^i(t) + \sum_{w \in \mathcal{N}_l} a_{l,w}^{i+1}(t)\mathbf{c}_w^*(t), \tag{6.1}$$

where $i$ denotes the number of iterations, $l$ denotes the follower agent, $\mathcal{N}_l$ represents the neighbor set of agent $l$, $\mathbf{c}_w^*(t)$ is the global coordinates of the leader agent and $\hat{\mathbf{c}}_q^i(t)$ is the estimated coordinates values of the follower agent. $a_{l,q}^{i+1}(t)$ and $a_{l,w}^{i+1}(t)$ are the calculated barycentric coordinate.

Consider all follower agents, one has a compact form,

$$\hat{\mathbf{C}}_s^{i+1}(t) = \mathbf{P}^{i+1}(t)\hat{\mathbf{C}}_s^{i}(t) + \mathbf{B}^{i+1}(t)\mathbf{C}_a^{*}(t), \tag{6.2}$$

where $\hat{\mathbf{C}}_s^{i}(t) = \left[\left(\hat{\mathbf{c}}_{m+1}^{i}(t)\right)^{\mathrm{T}}, \ldots, \left(\hat{\mathbf{c}}_{N}^{i}(t)\right)^{\mathrm{T}}\right]^{\mathrm{T}}$, and $\mathbf{C}_a^{*}(t) = \left[\left(\mathbf{c}_1^{*}(t)\right)^{\mathrm{T}}, \ldots, \left(\mathbf{c}_m^{*}(t)\right)^{\mathrm{T}}\right]^{\mathrm{T}}$.

$\mathbf{P}^{i+1}(t) = \begin{bmatrix} 0 & a_{m+1,m+2}^{i+1}(t) & \cdots & a_{m+1,N}^{i+1}(t) \\ a_{m+2,m+1}^{i+1}(t) & 0 & \cdots & a_{m+2,N}^{i+1}(t) \\ \vdots & \vdots & \ddots & \vdots \\ a_{N,m+1}^{i+1}(t) & a_{N,m+2}^{i+1}(t) & \cdots & 0 \end{bmatrix} \in \mathbb{R}^{(N-m)\times(N-m)}$ and the elements in $\mathbf{P}^{i+1}(t)$ are the calculated barycentric coordinates between followers.

$\mathbf{B}^{i+1}(t) = \begin{bmatrix} a_{m+1,1}^{i+1}(t) & a_{m+1,2}^{i+1}(t) & \cdots & a_{m+1,m}^{i+1}(t) \\ a_{m+2,1}^{i+1}(t) & a_{m+2,2}^{i+1}(t) & \cdots & a_{m+2,m}^{i+1}(t) \\ \vdots & \vdots & \ddots & \vdots \\ a_{N,1}^{i+1}(t) & a_{N,2}^{i+1}(t) & \cdots & a_{N,m}^{i+1}(t) \end{bmatrix} \in \mathbb{R}^{(N-m)\times m}$, and the elements in $\mathbf{B}^{i+1}(t)$ are the calculated barycentric coordinates between followers and leaders.

Under the assumption that the sensor is working in an ideal condition (i.e., the sensor operates normally, the measured information is accurate and there is no external interference), the accurate localization estimation can be guaranteed after applying (6.2) to the dynamic MASs. It should be pointed out that the ideal condition assumption may not be satisfied in practical applications. As mentioned in subsection A, the measured relative distance as the only available information to calculate the barycentric coordinates. Affected by the complex application environment, sensors are more likely to fail than other components in practice, and the performance of localization estimation also depends heavily on measurement information from sensors. It is well known that sensor failure result in measurement errors, which will add the difficulty in designing the estimation algorithm to guarantee the accurate localization.

This work investigates the accurate localization estimation of the dynamic MASs with sensor multiplicative failures. Based on the above analysis, it is foreseeable that some of the measured relative distance data in the considered multi-agent systems may deviate from the actual relative distance if the sensors experience failures. Some challenging and interesting questions naturally arise: Firstly, how to build the mathematical model describing the relationship between barycentric coordinates and relative distance only based on incomplete measurement information. Secondly, how to design the adaptive estimation algorithm to eliminate the influence of sensor failure on the final localization accuracy. Thirdly, only the relative distance is available. The model of the agent is unknown, which means that no state information of the agent can be obtained. Note that when applying the Kalman filter algorithm, both the state information of the agent and the sensor measurement data are required. Therefore, fusion estimation methods, such as the Kalman filter algorithm, cannot be applied to handle the problem of sensor failure.

The objective of this work is to realize the accurate localization of dynamic multi-agent systems with sensor multiplicative failures, that is, $\lim_{i\to\infty} \left\|\mathbf{E}_s^{i+1}(t)\right\|_\infty = 0$, where $\mathbf{E}_s^{i+1}(t) = \hat{\mathbf{C}}_s^{i+1}(t) - \mathbf{C}_s^*(t)$ represents the coordinates estimation error and $\mathbf{C}_s^*(t)$ is the real coordinates of all followers.

## 6.3 Data-Driven Adaptive Distributed Localization Under Sensor Multiplicative Failures

In this work, the sensor multiplicative failures is considered, the definition of which is given as follows.

**Definition 6.1** (*Sensor multiplicative failures*) [3, 4]: The sensor used to measure the system variable $\xi^F(t) \in \mathbb{R}$ can be said to be failed at the time $t_F$.

$$\xi^F(t) = \alpha(t)\xi(t), 0 < \alpha(t) \le 1, \forall t > t_F. \tag{6.3}$$

Based on Definition 6.1, sensor multiplicative failures in multi-agent systems can be described as follows

$$d_{l,x}^{i,F}(t) = \alpha_{l,x}^i(t) d_{l,x}^i(t), \tag{6.4}$$

where $0 < \alpha_{l,x}^i(t) \le 1$ represents the failure factor. $d_{l,x}^i(t)$ is the actual relative distance and $d_{l,x}^{i,F}(t)$ is the measured relative distance with sensor failure. $\alpha_{l,x}^i(t) = 1$ indicates that the sensor is working in good condition, while $\alpha_{l,x}^i(t) = 0$ indicates sensor failure, in which the sensor cannot measure the relative distance data.

**Assumption 6.1** In this work, only a part of the sensors are failure, there are still a few sensors that are operating normally. The sensor multiplicative failures do not occur in every iteration, there exists at least one iteration, the sensors can operate normally. The probability of sensor failure between adjacent iterations is independent of each other.

The measured relative distance data is used to calculate the barycentric coordinates. To eliminate the adverse effect of sensor failure on relative distance data, we first constructed a mathematical model to describe the relationship between barycentric coordinates and the relative distance. The details are as follows

$$a_{l,x}^{i+1}(t) = f\left(d_{l,x}^{i+1}(t)\right), \tag{6.5}$$

where $f(\cdot)$ is an unknown function.

Due to the calculation of the barycentric coordinates using the Cayley-Menger determinant, the unknown function $f(\cdot)$ is linear. Note that the expression of the function, which is used to describe the relationship between relative distance and barycentric coordinates, will be more complex when considering the higher dimensional space, so the simplified function $f(\cdot)$ is introduced to replace the complex linear expression.

**Remark 6.1** The barycentric coordinates is the ratio of the area between triangles, so the larger relative distance corresponds to the larger barycentric coordinates. Meanwhile, the relative distance is greater than zero. Therefore, $f(\cdot)$ is a monotonically increasing function, which satisfies the Lipschitz condition, i.e., $|f(d_1) - f(d_2)| \le \delta_d |d_1 - d_2|$, $\delta_d > 0$ is a bounded constant.

By virtue of Cauchy's mean value theorem [5, 6], one can obtain a linear data model to describe the relationship of barycentric coordinates between two consecutive iterations

$$\begin{aligned} & a_{l,x}^{i+1}(t) - a_{l,x}^{i}(t) \\ & = f\left(d_{l,x}^{i+1}(t)\right) - f\left(d_{l,x}^{i}(t)\right) \\ & = \frac{\partial f}{\partial d_{l,x}^{i+1}(t)}\left(d_{l,x}^{i+1}(t) - d_{l,x}^{i}(t)\right) \\ & = \theta_{l,x}^{i+1}(t)\Delta d_{l,x}^{i+1}(t), \end{aligned} \tag{6.6}$$

where $\theta_{l,x}^{i+1}(t)$ is an unknown parameter and $\Delta$ is a difference factor.

To compensate the relative distance data measured in case of sensor failure, based on the obtained linear data model (6.6), a data-driven adaptive localization estimation scheme is proposed as follows.

Firstly, the design of an adaptive parameter estimation algorithm. Since $\theta_{l,x}^{i+1}(t)$ is unknown, an estimation criterion function of $\theta_{l,x}^{i+1}(t)$ is designed below

$$\begin{aligned} J\left(\hat{\theta}_{l,x}^{i+1}(t)\right) = & \left|\Delta a_{l,x}^{i}(t) - \hat{\theta}_{l,x}^{i+1}(t)\Delta d_{l,x}^{i}(t)\right|^2 \\ & + \mu\left|\hat{\theta}_{l,x}^{i+1}(t) - \hat{\theta}_{l,x}^{i}(t)\right|^2, \end{aligned} \tag{6.7}$$

where $\mu > 0$ is a weighting factor.

Differentiating Eq. (6.7) with respect to $\hat{\theta}_{l,x}^{i+1}(t)$ and setting it to zero, one can obtain that

$$\begin{aligned} & -\Delta d_{l,x}^{i}(t)\left(\Delta a_{l,x}^{i}(t) - \hat{\theta}_{l,x}^{i+1}(t)\Delta d_{l,x}^{i}(t) + \hat{\theta}_{l,x}^{i}(t)\Delta d_{l,x}^{i}(t)\right. \\ & \left. - \hat{\theta}_{l,x}^{i}(t)\Delta d_{l,x}^{i}(t)\right) + \mu\left(\hat{\theta}_{l,x}^{i+1}(t) - \hat{\theta}_{l,x}^{i}(t)\right) = 0. \end{aligned} \tag{6.8}$$

By solving (6.8), we have

$$\begin{aligned} & \left(\left|\Delta d_{l,x}^{i}(t)\right|^2 + \mu\right)\left(\hat{\theta}_{l,x}^{i+1}(t) - \hat{\theta}_{l,x}^{i}(t)\right) = \\ & \Delta d_{l,x}^{i}(t)\left(\Delta a_{l,x}^{i}(t) - \hat{\theta}_{l,x}^{i}(t)\Delta d_{l,x}^{i}(t)\right). \end{aligned} \tag{6.9}$$

Then, the parameter estimation algorithm is given as

$$\hat{\theta}_{l,x}^{i+1}(t) = \hat{\theta}_{l,x}^{i}(t) + \frac{\eta \Delta d_{l,x}^{i}(t)}{\mu + \left|\Delta d_{l,x}^{i}(t)\right|^2} \left(\Delta a_{l,x}^{i}(t) - \hat{\theta}_{l,x}^{i}(t) \Delta d_{l,x}^{i}(t)\right), \tag{6.10}$$

where $\eta \in (0, 1)$ represents the step factor, which is added to make the estimation method more general and flexible.

Secondly, the design of data-driven adaptive relative distance compensation algorithm.

Let $\hat{d}_{l,x}^{1}(t) = d_{l,x}^{1,F}(t)$. For $i > 1$, the data-driven adaptive relative distance compensation algorithm is designed as

$$\hat{d}_{l,x}^{i+1}(t) = \hat{d}_{l,x}^{i}(t) + \frac{\rho \hat{\theta}_{l,x}^{i+1}(t)}{\lambda + \left|\hat{\theta}_{l,x}^{i+1}(t)\right|^2} \left(a_{l,x}^{i+1}(t) - \hat{a}_{l,x}^{i}(t)\right), \tag{6.11}$$

where $\lambda > 0$ is a weighting factor and $0 < \rho \leq 1$ represents the step factor, which is added to make the compensation algorithm more flexible. The barycentric coordinates $a_{l,x}^{i+1}(t)$ is calculated based on the measured relative distance $d_{l,x}^{i+1,F}(t)$.

After then, the estimated barycentric coordinates $\hat{a}_{l,x}^{i+1}(t)$ can be calculated based on the estimated relative distance $\hat{d}_{l,x}^{i+1}(t)$.

Based on the analysis from (6.7) to (6.11), the whole data-driven adaptive distributed localization estimation scheme is shown below

$$\hat{\theta}_{l,x}^{i+1}(t) = \hat{\theta}_{l,x}^{i}(t) + \frac{\eta \Delta d_{l,x}^{i}(t)}{\mu + \left|\Delta d_{l,x}^{i}(t)\right|^2} \left(\Delta a_{l,x}^{i}(t) - \hat{\theta}_{l,x}^{i}(t) \Delta d_{l,x}^{i}(t)\right). \tag{6.12}$$

$$\hat{d}_{l,x}^{i+1}(t) = \hat{d}_{l,x}^{i}(t) + \frac{\rho \hat{\theta}_{l,x}^{i+1}(t)}{\lambda + \left|\hat{\theta}_{l,x}^{i+1}(t)\right|^2} \left(a_{l,x}^{i+1}(t) - \hat{a}_{l,x}^{i}(t)\right). \tag{6.13}$$

$$\hat{\mathbf{c}}_{l}^{i+1}(t) = \sum_{q \in \mathcal{N}_l} \hat{a}_{l,q}^{i+1}(t) \hat{\mathbf{c}}_{q}^{i}(t) + \sum_{w \in \mathcal{N}_l} \hat{a}_{l,w}^{i+1}(t) \mathbf{c}_{w}^{*}(t). \tag{6.14}$$

A block diagram of the proposed localization estimation scheme is provided to describe the processing of the measured relative distance data, as depicted in Fig. 6.1, where $\tau^{-1}$ represents an iteration backward operator, i.e., $\tau^{-1} \hat{a}_{l,x}^{i+1}(t) = \hat{a}_{l,x}^{i}(t)$.

The process of data-driven adaptive distributed localization estimation algorithm is presented in Algorithm 6.1.

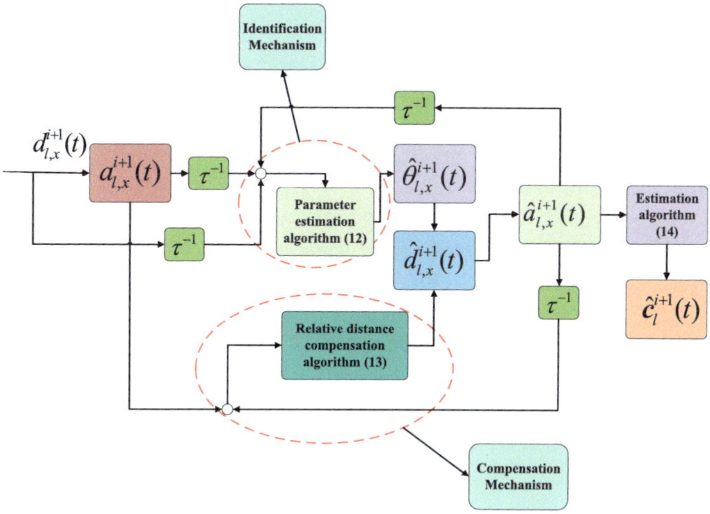


**Fig. 6.1** The block diagram of the proposed adaptive localization estimation algorithm

---

**Algorithm 6. 1**. Estimate the real coordinate of the follower agents under sensor multiplicative failures.

---

**for** $i = 1 : Iteration.$

**initialization** Given the initial iteration coordinate estimation value $\hat{\mathbf{c}}_l^i(t)$ and the initial estimation parameter $\hat{\theta}_{l,x}^i(t)$ of the follower agents.

**Step 1.** Each agent transmits its own estimated coordinates $\hat{\mathbf{c}}_l^i(t)$ to its neighbors.

**Step 2.** Measure the relative distance, and calculate the barycentric coordinate.

**Step 3.** Identify the estimation parameter $\hat{\theta}_{l,x}^{i+1}(t)$ based on measured relative distance and calculate the barycentric coordinate:

$$\hat{\theta}_{l,x}^{i+1}(t) = \hat{\theta}_{l,x}^{i}(t) + \frac{\eta \Delta d_{l,x}^{i}(t)}{\mu + \left|\Delta d_{l,x}^{i}(t)\right|^2}\left(\Delta a_{l,x}^{i}(t) - \hat{\theta}_{l,x}^{i}(t)\Delta d_{l,x}^{i}(t)\right).$$

**Step 4.** Compensate the relative distance data is based on the identified parameter $\hat{\theta}_{l,x}^{i+1}(t)$:

$$\hat{d}_{l,x}^{i+1}(t) = \hat{d}_{l,x}^{i}(t) + \frac{\rho \hat{\theta}_{l,x}^{i+1}(t)}{\lambda + \left|\hat{\theta}_{l,x}^{i+1}(t)\right|^2}\left(a_{l,x}^{i+1}(t) - \hat{a}_{l,x}^{i}(t)\right).$$

**Step 5.** Calculate the estimated barycentric coordinates $\hat{a}_{l,x}^{i+1}(t)$ based on the estimated relative distance $\hat{d}_{l,x}^{i+1}(t)$.

**Step 6.** Update

$$\hat{\mathbf{c}}_l^{i+1}(t) = \sum_{q \in \mathcal{N}_l} \hat{a}_{l,q}^{i+1}(t)\hat{\mathbf{c}}_q^i(t) + \sum_{w \in \mathcal{N}_l} \hat{a}_{l,w}^{i+1}(t)\mathbf{c}_w^*(t).$$

**Step 7.** Update $i = i + 1$ and return to **Step 1**.

**end**

---

## 6.4 Convergence Analysis

This section proves the convergence of the data-driven adaptive distributed localization estimation scheme in the case of sensor multiplicative failures.

**Theorem 6.1** *Consider the localization estimation of dynamic MASs with repetitive characteristics. The proposed data-driven adaptive distributed localization estimation scheme (6.12)–(6.14) assures accurate real-time localization under sensor multiplicative failures, i.e.,* $\lim_{i\to\infty} \left\|\mathbf{E}_s^{i+1}(t)\right\|_\infty = 0$.

***Proof*** The proof is divided into three parts: (i) the boundedness of the estimation parameter; (ii) the convergence of the barycentric coordinates estimation error; (iii) the convergence of the localization estimation error.

**Part (i)** The boundedness of the estimation parameter $\hat{\theta}_{l,x}^{i+1}(t)$.

Define the parameter estimation error as $\tilde{\theta}_{l,x}^{i+1}(t) = \hat{\theta}_{l,x}^{i+1}(t) - \theta_{l,x}^{i+1}(t)$, where $\theta_{l,x}^{i+1}(t)$ is the actual parameter.

Subtract $\theta_{l,x}^{i+1}(t)$ from both sides of (6.12), one has

$$\begin{aligned}
\hat{\theta}_{l,x}^{i+1}(t) - \theta_{l,x}^{i+1}(t) &= \hat{\theta}_{l,x}^{i}(t) - \theta_{l,x}^{i}(t) + \frac{\eta \Delta d_{l,x}^{i}(t)}{\mu + \left|\Delta d_{l,x}^{i}(t)\right|^2} \\
&\times \left(\Delta a_{l,x}^{i}(t) - \hat{\theta}_{l,x}^{i}(t)\Delta d_{l,x}^{i}(t)\right) + \theta_{l,x}^{i}(t) - \theta_{l,x}^{i+1}(t) \\
\tilde{\theta}_{l,x}^{i+1}(t) &= \tilde{\theta}_{l,x}^{i}(t) + \frac{\eta \Delta d_{l,x}^{i}(t)}{\mu + \left|\Delta d_{l,x}^{i}(t)\right|^2}\left(\theta_{l,x}^{i}(t)\Delta d_{l,x}^{i}(t)\right. \\
&\left. -\hat{\theta}_{l,x}^{i}(t)\Delta d_{l,x}^{i}(t)\right) + \theta_{l,x}^{i}(t) - \theta_{l,x}^{i+1}(t) \\
\tilde{\theta}_{l,x}^{i+1}(t) &= \left(1 - \frac{\eta\left|\Delta d_{l,x}^{i}(t)\right|^2}{\mu + \left|\Delta d_{l,x}^{i}(t)\right|^2}\right)\tilde{\theta}_{l,x}^{i}(t) + \theta_{l,x}^{i}(t) - \theta_{l,x}^{i+1}(t).
\end{aligned} \tag{6.15}$$

Take absolute value on both sides of (6.15) yields

$$\left|\tilde{\theta}_{l,x}^{i+1}(t)\right| \le \left|1 - \frac{\eta\left|\Delta d_{l,x}^{i}(t)\right|^2}{\mu + \left|\Delta d_{l,x}^{i}(t)\right|^2}\right|\left|\tilde{\theta}_{l,x}^{i}(t)\right| + \left|\theta_{l,x}^{i}(t) - \theta_{l,x}^{i+1}(t)\right|. \tag{6.16}$$

Since $\eta \in (0, 1)$ and $\mu > 0$, one can ensure that $0 < \frac{\eta\left|\Delta d_{l,x}^{i}(t)\right|^2}{\mu + \left|\Delta d_{l,x}^{i}(t)\right|^2} < 1$ by selecting the appropriate $\eta$ and $\mu$. After then, there must exist a constant $\sigma_1$ such that

$$0 < \left|1 - \frac{\eta\left|\Delta d_{l,x}^{i}(t)\right|^2}{\mu + \left|\Delta d_{l,x}^{i}(t)\right|^2}\right| = \sigma_1 < 1. \tag{6.17}$$

According to the analysis of Remark 6.1, $\left|\theta_{l,x}^{i}(t)\right| < \delta_d$ is bounded, then one has $\left|\theta_{l,x}^{i}(t) - \theta_{l,x}^{i+1}(t)\right| \le 2\delta_d$ is bounded. Based on the above analysis (6.15)–(6.17), (6.16) becomes

$$\left|\tilde{\theta}_{l,x}^{i+1}(t)\right| \leq \sigma_1 \left|\tilde{\theta}_{l,x}^{i}(t)\right| + 2\delta_d. \tag{6.18}$$

By solving (6.18), one can obtain that

$$\left|\tilde{\theta}_{l,x}^{i+1}(t)\right| \leq \sigma_1^{i+1} \left|\tilde{\theta}_{l,x}^{0}(t)\right| + \frac{2\delta_d}{1-\sigma_1}. \tag{6.19}$$

Since $\tilde{\theta}_{l,x}^{0}(t)$ is bounded and $0 < \sigma_1 < 1$, (6.19) implies that $\tilde{\theta}_{l,x}^{i+1}(t)$ is bounded. $\theta_{l,x}^{i+1}(t)$ is also bounded, then the boundedness of $\hat{\theta}_{l,x}^{i+1}(t)$ is proved.

**Part (ii)** The convergence of the estimation error $\tilde{a}_{l,x}^{i+1}(t) = a_{l,x}^{*}(t) - \hat{a}_{l,x}^{i+1}(t)$, where $a_{l,x}^{*}(t)$ is the actual barycentric coordinates.

Based on obtained linear data model (6.6) and compensation algorithm (6.13), the estimation error can be expressed as

$$\begin{aligned}
\tilde{a}_{l,x}^{i+1}(t) &= a_{l,x}^{*}(t) - \hat{a}_{l,x}^{i+1}(t) \\
&= a_{l,x}^{*}(t) - \hat{a}_{l,x}^{i}(t) - \hat{\theta}_{l,x}^{i+1}(t)\Delta\hat{d}_{l,x}^{i+1}(t) \\
&= \tilde{a}_{l,x}^{i}(t) - \frac{\rho\left|\hat{\theta}_{l,x}^{i+1}(t)\right|^2}{\lambda+\left|\hat{\theta}_{l,x}^{i+1}(t)\right|^2}\left(a_{l,x}^{i+1}(t) - a_{l,x}^{*}(t) + a_{l,x}^{*}(t) - \hat{a}_{l,x}^{i}(t)\right) \\
&= \left(1 - \frac{\rho\left|\hat{\theta}_{l,x}^{i+1}(t)\right|^2}{\lambda+\left|\hat{\theta}_{l,x}^{i+1}(t)\right|^2}\right)\tilde{a}_{l,x}^{i}(t) + \frac{\rho\left|\hat{\theta}_{l,x}^{i+1}(t)\right|^2}{\lambda+\left|\hat{\theta}_{l,x}^{i+1}(t)\right|^2}\left(a_{l,x}^{*}(t) - a_{l,x}^{i+1}(t)\right).
\end{aligned} \tag{6.20}$$

Take absolute value on both sides of (6.20), one has

$$\left|\tilde{a}_{l,x}^{i+1}(t)\right| \leq \left|\left(1 - \frac{\rho\left|\hat{\theta}_{l,x}^{i+1}(t)\right|^2}{\lambda+\left|\hat{\theta}_{l,x}^{i+1}(t)\right|^2}\right)\right|\left|\tilde{a}_{l,x}^{i}(t)\right| + \left|\frac{\rho\left|\hat{\theta}_{l,x}^{i+1}(t)\right|^2}{\lambda+\left|\hat{\theta}_{l,x}^{i+1}(t)\right|^2}\right|\left|\left(a_{l,x}^{*}(t) - a_{l,x}^{i+1}(t)\right)\right|. \tag{6.21}$$

Since $0 < \rho \leq 1$ and $\lambda > 0$, one can ensure that $0 < \frac{\rho\left|\hat{\theta}_{l,x}^{i+1}(t)\right|^2}{\lambda+\left|\hat{\theta}_{l,x}^{i+1}(t)\right|^2} < 1$ by selecting the appropriate $\rho$ and $\lambda$. Then, one has $\left|\left(1 - \frac{\rho\left|\hat{\theta}_{l,x}^{i+1}(t)\right|^2}{\lambda+\left|\hat{\theta}_{l,x}^{i+1}(t)\right|^2}\right)\right| = \sigma_2 < 1$, (6.21) becomes

$$\left|\tilde{a}_{l,x}^{i+1}(t)\right| \leq \sigma_2\left|\tilde{a}_{l,x}^{i}(t)\right| + \left|\frac{\rho\left|\hat{\theta}_{l,x}^{i+1}(t)\right|^2}{\lambda+\left|\hat{\theta}_{l,x}^{i+1}(t)\right|^2}\right|\left|\left(a_{l,x}^{*}(t) - a_{l,x}^{i+1}(t)\right)\right|. \tag{6.22}$$

Note that $a_{l,x}^{i+1}(t)$ is calculated based on the measured relative distance. Consequently, both $a_{l,x}^{*}(t)$ and $a_{l,x}^{i+1}(t)$ are bounded. According to the above analysis, $\left|\frac{\rho\left|\hat{\theta}_{l,x}^{i+1}(t)\right|^2}{\lambda+\left|\hat{\theta}_{l,x}^{i+1}(t)\right|^2}\right|\left|\left(a_{l,x}^{*}(t) - a_{l,x}^{i+1}(t)\right)\right| = \sigma_3$ is bounded. Since $\lambda$ and $\sigma_3$ are independent, selecting the appropriate weighting factor $\lambda$ can guarantee that $\sigma_3$ converge to the acceptable small bounded value.

By solving (6.22), one has

$$\left|\tilde{a}_{l,x}^{i+1}(t)\right| \leq \sigma_2 \left|\tilde{a}_{l,x}^{i}(t)\right| + \sigma_3 \leq \cdots \leq \sigma_2^{i+1} \left|\tilde{a}_{l,x}^{0}(t)\right| + \tfrac{\sigma_3}{1-\sigma_2}, \tag{6.23}$$

which implies that $\tilde{a}_{l,x}^{i+1}(t)$ is bounded, that is to say, the estimated barycentric coordinates $\hat{a}_{l,x}^{i+1}(t)$ can converge to $a_{l,x}^{*}(t)$ with a small acceptable error when the iteration goes to infinity.

**Part (iii)** The convergence of the localization estimation error $\mathbf{E}_s^{i+1}(t)$.

Consider all $N-m$ followers, one can obtain that

$$\hat{\mathbf{C}}_s^{i+1}(t) = \hat{\mathbf{P}}^{i+1}(t)\hat{\mathbf{C}}_s^{i}(t) + \hat{\mathbf{B}}^{i+1}(t)\mathbf{C}_a^{*}(t), \tag{6.24}$$

where $\hat{\mathbf{P}}^{i+1}(t) = \begin{bmatrix} 0 & \hat{a}_{m+1,m+2}^{i+1}(t) & \cdots & \hat{a}_{m+1,N}^{i+1}(t) \\ \hat{a}_{m+2,m+1}^{i+1}(t) & 0 & \cdots & \hat{a}_{m+2,N}^{i+1}(t) \\ \vdots & \vdots & \ddots & \vdots \\ \hat{a}_{N,m+1}^{i+1}(t) & \hat{a}_{N,m+2}^{i+1}(t) & \cdots & 0 \end{bmatrix} \in \mathbb{R}^{(N-m)\times(N-m)}$ characterizes the information interaction between the followers and is the estimates of the actual barycentric coordinate matrix $\mathbf{P}^{i+1}(t)$.

$\hat{\mathbf{B}}^{i+1}(t) = \begin{bmatrix} \hat{a}_{m+1,1}^{i+1}(t) & \hat{a}_{m+1,2}^{i+1}(t) & \cdots & \hat{a}_{m+1,m}^{i+1}(t) \\ \hat{a}_{m+2,1}^{i+1}(t) & \hat{a}_{m+2,2}^{i+1}(t) & \cdots & \hat{a}_{m+2,m}^{i+1}(t) \\ \vdots & \vdots & \ddots & \vdots \\ \hat{a}_{N,1}^{i+1}(t) & \hat{a}_{N,2}^{i+1}(t) & \cdots & \hat{a}_{N,m}^{i+1}(t) \end{bmatrix} \in \mathbb{R}^{(N-m)\times m}$ represents the relative position between followers and leaders, and is the estimates of the actual barycentric coordinate matrix $\mathbf{B}^{i+1}(t)$.

Let $\tilde{\mathbf{P}}^{i+1}(t) = \hat{\mathbf{P}}^{i+1}(t) - \mathbf{P}^{i+1}(t)$ and $\tilde{\mathbf{B}}^{i+1}(t) = \hat{\mathbf{B}}^{i+1}(t) - \mathbf{B}^{i+1}(t)$. Then, (6.24) can be expressed as

$$\hat{\mathbf{C}}_s^{i+1}(t) = \tilde{\mathbf{P}}^{i+1}(t)\hat{\mathbf{C}}_s^{i}(t) + \tilde{\mathbf{B}}^{i+1}(t)\mathbf{C}_a^{*}(t) + \mathbf{P}^{i+1}(t)\hat{\mathbf{C}}_s^{i}(t) + \mathbf{B}^{i+1}(t)\mathbf{C}_a^{*}(t). \tag{6.25}$$

Subtract $\mathbf{C}_s^{*}(t)$ from both sides of Eq. (6.25), we have

$$\begin{aligned} \mathbf{E}_s^{i+1}(t) = \mathbf{P}^{i+1}(t)\mathbf{E}_s^{i}(t) + \tilde{\mathbf{P}}^{i+1}(t)\mathbf{E}_s^{i}(t) \\ +\tilde{\mathbf{P}}^{i+1}(t)\mathbf{C}_s^{*}(t) + \tilde{\mathbf{B}}^{i+1}(t)\mathbf{C}_a^{*}(t) + \\ \left[\mathbf{B}^{i+1}(t)\mathbf{C}_a^{*}(t) + \left(\mathbf{P}^{i+1}(t) - \mathbf{I}\right)\mathbf{C}_s^{*}(t)\right]. \end{aligned} \tag{6.26}$$

The fifth term $\mathbf{B}^{i+1}(t)\mathbf{C}_a^{*}(t) + \left(\mathbf{P}^{i+1}(t) - \mathbf{I}\right)\mathbf{C}_s^{*}(t)$ of (6.26) is expressed as

$$
\begin{aligned}
&\mathbf{B}^{i+1}(t)\mathbf{C}_a^*(t) + \left(\mathbf{P}^{i+1}(t) - \mathbf{I}\right)\mathbf{C}_s^*(t) = \\
&\begin{bmatrix}
a_{m+1,1}^{i+1}(t) & a_{m+1,2}^{i+1}(t) & \cdots & a_{m+1,m}^{i+1}(t) \\
a_{m+2,1}^{i+1}(t) & a_{m+2,2}^{i+1}(t) & \cdots & a_{m+2,m}^{i+1}(t) \\
\vdots & \vdots & \ddots & \vdots \\
a_{N,1}^{i+1}(t) & a_{N,2}^{i+1}(t) & \cdots & a_{N,m}^{i+1}(t)
\end{bmatrix}
\begin{bmatrix}
\mathbf{c}_1^*(t) \\ \mathbf{c}_2^*(t) \\ \vdots \\ \mathbf{c}_m^*(t)
\end{bmatrix} \\
&+ \begin{bmatrix}
-1 & a_{m+1,m+2}^{i+1}(t) & \cdots & a_{m+1,N}^{i+1}(t) \\
a_{m+2,m+1}^{i+1}(t) & -1 & \cdots & a_{m+2,N}^{i+1}(t) \\
\vdots & \vdots & \ddots & \vdots \\
a_{N,m+1}^{i+1}(t) & a_{N,m+2}^{i+1}(t) & \cdots & -1
\end{bmatrix}
\begin{bmatrix}
\mathbf{c}_{m+1}^*(t) \\ \mathbf{c}_{m+2}^*(t) \\ \vdots \\ \mathbf{c}_N^*(t)
\end{bmatrix}.
\end{aligned}
\tag{6.27}
$$

One can know that $\mathbf{c}_l(t) = a_{lh}(t)\mathbf{c}_h(t) + a_{lj}(t)\mathbf{c}_j(t) + a_{lk}(t)\mathbf{c}_k(t)$ and $\mathbf{c}_l^*(t)$ is the real coordinates. Due to the fact that when the real coordinates of the neighboring agents are known, the real coordinates of the agent to be located can be uniquely determined, one has

$$
\begin{aligned}
a_{m+1,1}^{i+1}(t)\mathbf{c}_1^*(t) + \cdots + a_{m+1,m}^{i+1}(t)\mathbf{c}_m^*(t) + a_{m+1,m+2}^{i+1}(t) \\
\times \mathbf{c}_{m+2}^*(t) + \cdots + a_{m+1,N}^{i+1}(t)\mathbf{c}_N^*(t) = \mathbf{c}_{m+1}^*(t), \\
\vdots \\
a_{N,1}^{i+1}(t)\mathbf{c}_1^*(t) + \cdots + a_{N,m}^{i+1}(t)\mathbf{c}_m^*(t) + a_{N,m+1}^{i+1}(t)\mathbf{c}_{m+1}^*(t) \\
+ \cdots + a_{N,N-1}^{i+1}(t)\mathbf{c}_{N-1}^*(t) = \mathbf{c}_N^*(t).
\end{aligned}
\tag{6.28}
$$

Then, $\mathbf{B}^{i+1}(t)\mathbf{C}_a^*(t) + \left(\mathbf{P}^{i+1}(t) - \mathbf{I}\right)\mathbf{C}_s^*(t) = \mathbf{0}$ can be obtained.
According to the analysis from (6.27) to (6.28), (6.26) becomes

$$
\mathbf{E}_s^{i+1}(t) = \mathbf{P}^{i+1}(t)\mathbf{E}_s^i(t) + \tilde{\mathbf{P}}^{i+1}(t)\mathbf{E}_s^i(t) + \tilde{\mathbf{P}}^{i+1}(t)\mathbf{C}_s^*(t) + \tilde{\mathbf{B}}^{i+1}(t)\mathbf{C}_a^*(t). \tag{6.29}
$$

The infinite norm of Eq. (6.29) is shown as follows,

$$
\begin{aligned}
\left\|\mathbf{E}_s^{i+1}(t)\right\|_\infty \le \left(\left\|\mathbf{P}^{i+1}(t)\right\|_\infty + \left\|\tilde{\mathbf{P}}^{i+1}(t)\right\|_\infty\right)\left\|\mathbf{E}_s^i(t)\right\|_\infty \\
+ \left\|\tilde{\mathbf{P}}^{i+1}(t)\right\|_\infty \left\|\mathbf{C}_s^*(t)\right\|_\infty + \left\|\tilde{\mathbf{B}}^{i+1}(t)\right\|_\infty \left\|\mathbf{C}_a^*(t)\right\|_\infty.
\end{aligned}
\tag{6.30}
$$

The element of $\mathbf{P}_{i+1}(t)$ represents the relative position between followers, so the sum of elements in each row is less than one. Based on the definition of infinite norm, $\left\|\mathbf{P}^{i+1}(t)\right\|_\infty = \sigma_4 < 1$ can be guaranteed.

Based on the analysis of part (ii), we can know that for a large iteration number $i_d$, $\left\|\tilde{\mathbf{P}}^{i_d+1}(t)\right\|_\infty \le \zeta$, $\left\|\tilde{\mathbf{B}}^{i_d+1}(t)\right\|_\infty \le \zeta$ is established, where $\zeta < 1 - \sigma_4$.

**Remark 6.2** According to the convergence analysis of part (ii), the estimated barycentric coordinates $\hat{a}_{l,q}^{i+1}(t)$ and $\hat{a}_{l,w}^{i+1}(t)$ can converge to the actual $a_{l,q}^{i+1}(t)$ and

$a_{l,w}^{i+1}(t)$. In other words, $\left\|\tilde{\mathbf{P}}^{i+1}(t)\right\|_{\infty}$ can converge to a small bounded value $\zeta$ when the number of iterations is large enough, then $\left\|\mathbf{P}^{i+1}(t)\right\|_{\infty}+\left\|\tilde{\mathbf{P}}^{i+1}(t)\right\|_{\infty}<1$ holds.

Since the real coordinates are bounded, (6.30) becomes

$$\left\|\mathbf{E}_{s}^{i+1}(t)\right\|_{\infty} \leq \sigma_{5}\left\|\mathbf{E}_{s}^{i}(t)\right\|_{\infty}+\zeta\sigma_{6}, \tag{6.31}$$

where $\sigma_{5}=\sigma_{4}+\zeta<1, \sigma_{6}=\left\|\mathbf{C}_{s}^{*}(t)\right\|_{\infty}+\left\|\mathbf{C}_{a}^{*}(t)\right\|_{\infty}<\infty$.

From (6.31), we have

$$\left\|\mathbf{E}_{s}^{i+1}(t)\right\|_{\infty} \leq \sigma_{5}^{i-i_{b}+1}\left\|\mathbf{E}_{s}^{i_{b}}(t)\right\|_{\infty}+\zeta\frac{\sigma_{6}}{1-\sigma_{5}}. \tag{6.32}$$

Since $0<\sigma_{5}<1$, one can obtain that

$$\lim_{i\to\infty}\sigma_{5}^{i-i_{b}+1}\left\|\mathbf{E}_{s}^{i_{b}}(t)\right\|_{\infty}=0. \tag{6.33}$$

Similarly, one can obtain that $\frac{\sigma_6}{1-\sigma_5}=\sigma_7<\infty$ is bounded. And $\sigma_7$ and $\zeta$ are independent, then one can obtain that

$$\limsup_{i\to\infty}\left\|\mathbf{E}_{s}^{i+1}(t)\right\|_{\infty} \leq \zeta\sigma_{7}. \tag{6.34}$$

It can be found that (6.34) is established for any $\zeta$, based on the analysis in Remark 6.2, we know that $\zeta$ can converge to zero when iteration goes to infinity. Then, one can obtain that $\lim_{i\to\infty}\left\|\mathbf{E}_{s}^{i+1}(t)\right\|_{\infty}=0$.

The proof of Theorem 6.1 is completed.

## 6.5 Numerical Simulation

In this section, we provide the simulation example to demonstrate the validity of the proposed data-driven adaptive distributed localization estimation scheme (6.12)–(6.14).

Consider a multi-agent system containing seven agents that perform circular motion in two-dimensional plane, where three agents are leaders, and the remaining four agents are followers. Figure 6.2 shows the real curves of the seven agents, and the arrow denotes the motion direction. The adjacent sets of the follower agents are given below: $N_{f_1}=\{l_3, f_3, f_4\}$, $N_{f_2}=\{l_2, l_3, f_1\}$, $N_{f_3}=\{l_1, l_2, f_2\}$, $N_{f_4}=\{l_1, l_3, f_1\}$, where $l$ denotes the leader agent and $f$ represents the follower agent.

Assume that the sensors of follower 2 and leader 2 experience multiplicative failure, that is the measured relative-distance $d_{f_2,l_2}(t)$, $d_{f_2,l_3}(t)$, $d_{f_2,f_1}(t)$ and $d_{l_2,f_3}(t)$ may deviate from the actual value. Set the failure factor $\alpha_{l,x}^{i}(t)$ varying randomly in the range of [0.6, 1] along the iteration axis.

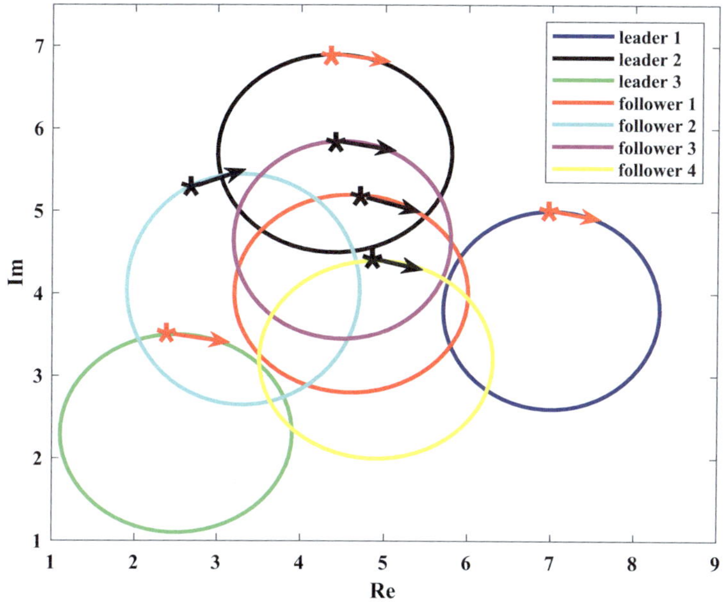


**Fig. 6.2** The real trajectories of follower and leader agents

The initial value of the estimation parameters and the estimation coordinates of the followers are set randomly. Select the constant factors as: $\eta_1 = 0.04$, $\eta_2 = 0.04$, $\eta_3 = 0.04$, $\eta_4 = 0.09$; $\mu_1 = 0.1$, $\mu_2 = 0.1$, $\mu_3 = 0.1$, $\mu_4 = 0.1$; $\rho_1 = 0.01$, $\rho_2 = 0.01$, $\rho_3 = 0.01$, $\rho_4 = 0.01$; $\lambda_1 = 0.5$, $\lambda_2 = 0.5$, $\lambda_3 = 0.5$, $\lambda_4 = 0.02$. Applying the proposed data-driven adaptive distributed localization estimation scheme (6.12)–(6.14), the simulation results are shown in Figs. 6.3 and 6.4.

Figure 6.3 shows the maximum estimation error $e_{l,\max}(i) = \max_t \left\{ \left\| \hat{\mathbf{c}}_l^i(t) - \mathbf{c}_l^*(t) \right\| \right\}$ of the four follower agents. One can find that that the estimation errors converge to zero at the 40th iteration. At the same time, the estimation error curves under the ideal environment are provided for comparative analysis, as shown in Fig. 6.4. The convergence rate in the sensor failure case is slower than in the ideal environment, but the proposed data-driven adaptive distributed localization estimation scheme (6.12)–(6.14) guarantees gradual convergence.

From the simulation and experimental results, one can conclude that: (i) The accurate localization of dynamic MASs with sensor multiplicative failures is achieved after applying (6.12)–(6.14). (ii) The proposed data-driven adaptive distributed localization scheme can eliminate the effects of sensor multiplicative failures on the localization estimation accuracy.

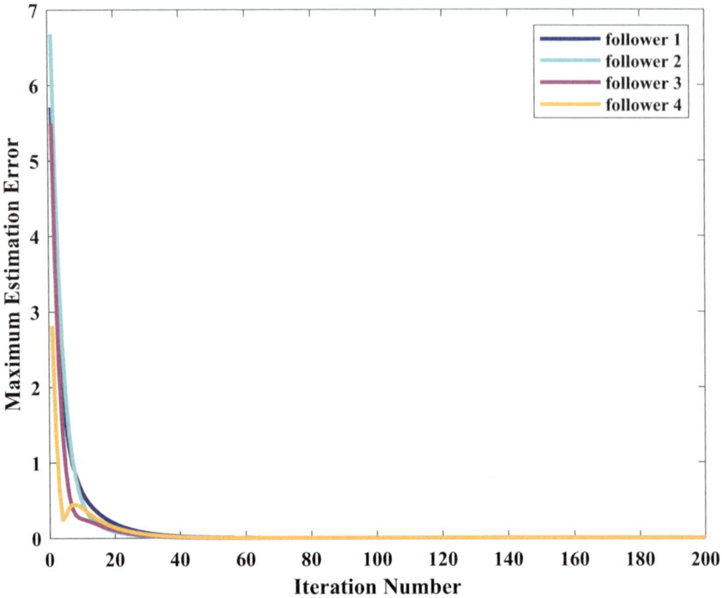


**Fig. 6.3** The maximum estimation error under sensor failures

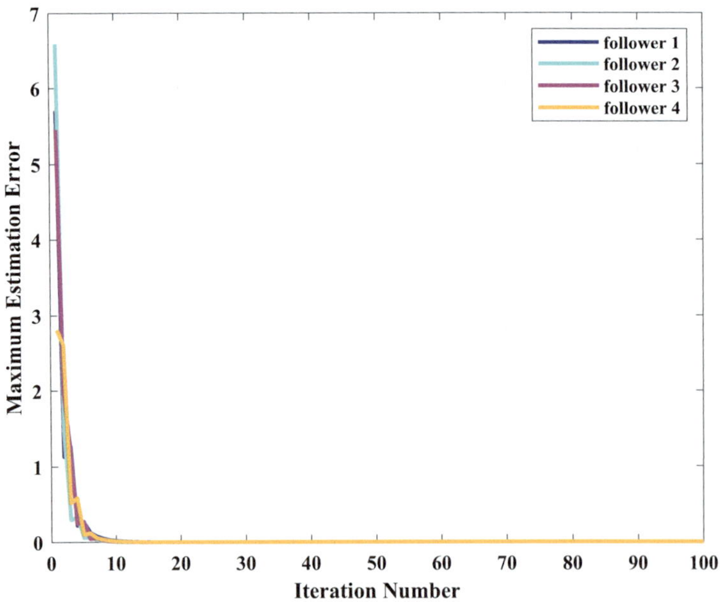


**Fig. 6.4** The maximum estimation error under ideal environment

## 6.6 Conclusions

This chapter considers the localization estimation of dynamic MASs with sensor multiplicative failures. A linear data model is constructed to characterize the relationship between barycentric coordinates and relative distance. A data-driven adaptive parameter estimation algorithm is designed based on the linear data model. After then, based on iterative learning, a data-driven adaptive distributed localization estimation scheme with a compensation mechanism is proposed, which guarantees accurate localization in the case of sensor multiplicative failures. The illustrative examples also prove the validity of the proposed localization methods. The localization in the case of sensor failure is considered in this work, the integration of localization estimation and formation control is of our future research interest.

## References


1. Li, H., Gao, Y., Shi, P., et al.: Observer-based fault detection for nonlinear systems with sensor fault and limited communication capacity. IEEE Trans. Autom. Control **61**(9), 2745–2751 (2016)
2. Dong, J., Yang, G.-H.: Reliable state feedback control of t-s fuzzy systems with sensor faults. IEEE Trans. Fuzzy Syst. **23**(2), 421–433 (2015)
3. Wu, H., Zhang, H.: Reliable mixed fuzzy static output feedback control for nonlinear systems with sensor faults. Automatica **41**, 1925–1932 (2005)
4. Rudin, W.: Principles of Mathematical Analysis. McGraw-Hill, New York (1976)
5. Hou, Z., Jin, S.: Model Free Adaptive Control: Theory and Applications. CRC Press (2013)
6. Goldenberg, D.K., Bihler, P., Yang, Y.R., et al.: Localization in sparse networks using sweeps. In: Proceedings of the 12th Annual International Conference on Mobile Computing and Networking, pp. 1–16 (2006)

# Chapter 7
# Distributed Sensor Network Localization Based on Local Bearing Measurement

This chapter solves the localization estimation of dynamic multi-agent systems (MASs) with sensor multiplicative failures, which is more general yet challenging to address than static sensor networks with ideal conditions. Barycentric coordinate is introduced to characterize the relative positions between agents. A new linear data model is constructed to represent the relationship between barycentric coordinates and relative distance. Based on the linear model, an adaptive parameter estimation algorithm is designed, and then it is applied to solve the relative distance compensation problem of the MASs with sensor multiplicative failures. Using the estimated parameter, a data-driven adaptive distributed localization estimation scheme based on iterative learning is proposed, in which only the measured relative distance data is available instead of the system model information. Key to obtaining accurate localization is overcoming the difficulties from inaccurate relative distance variables due to sensor failure via the data-driven adaptive relative distance compensation method. The numerical examples verify the effectiveness of the proposed methods.

## 7.1 Introduction

Network localization is the premise for multi-agent systems to perform complex tasks in many applications. In the past few decades, network localization algorithms have developed rapidly. The bearing-based localization algorithm receives growing interest due to the rapid development of bearings only sensors such as optical cameras, and sensor arrays.

Compared with the traditional distance-based localization estimation method, the advantage of the bearing-based one is that it can be expressed as a linear dynamic system and its global stability can be analyzed. A key technical difficulty for sensor network localization is that each sensor node has its own local reference frame, which is not aligned with global reference frame and the true orientation is not

 
Y. Lv et al., *Distributed Localization and Formation Control of Multi-Agent Systems with Local Sensing Information*, https://doi.org/10.1007/978-981-96-6015-5_7

known. Therefore, the orientation estimation problem, in the case of sensor node has no knowledge of global reference frame, deserves more attention. Furthermore, the measurement model with a directed graph is more general as relative measurement is often unidirectional in practice.

The contributions of this work are twofold. First, a new orientation estimation algorithm merely using the local bearing measurements is proposed. The relative orientation measurements are not need. The orientation can be determined indirectly using the rotation matrix which is calculated by estimating the unit direction vector between adjacent nodes. Second, a distributed network localization algorithm with a directed acyclic structure is proposed, which can be implemented and extended due to the cascade structure. The proposed algorithm has a concise form without orthogonalization procedure and meanwhile guarantees the global convergence.

The rest of this chapter is organized as follows. Section 7.2 shows the problem formulation for localization estimation. In Sect. 7.3, the distributed estimation algorithm and the convergence analysis is provided. Simulation results are presented in Sect. 7.4 and Conclusions are given in Sect. 7.5.

## 7.2 Problem Formulation

Consider an $n$-node sensor network in two-dimensional Euclidean space, including two anchor nodes (labelled as 1 and 2), whose positions in ${}^g\sum$ are known, and the remaining $(n-2)$ nodes are free nodes (labelled as $3, \ldots, n$) to be estimated in ${}^g\sum$.

In this work, the free node has its own local reference frame, which is not aligned with the global reference frame ${}^g\sum$, as shown in Fig.7.1. Suppose each free node can measure the bearing information of its neighbors under its local reference frame ${}^i\sum$ and receive the position information sent by the neighbors. In this work, we have two assumptions for the considered sensor network.

**Assumption 7.1** The positions of the sensor nodes do not overlap each other and the three neighbor nodes are not collinear.

**Assumption 7.2** The directed topology graph $\mathcal{G} = (v, \varepsilon)$ satisfies the Henneberg construction.

**Remark 7.1** Assumption 7.2 means that the free node $i \in \{3, \ldots, n\}$ receives the position information transmitted by the neighbors, and simultaneously sends its own position estimation information to other neighbors. Since the communication topology of the sensor networks satisfies the Henneberg construction and the communication between the sensor nodes is unidirectional, the neighbor node that transmits the position information to $i$ and the neighbor node that receives the information sent from $i$ are different nodes.

For the sensor network with directed acyclic leader-follower structure topology graph $\mathcal{G} = (v, \varepsilon)$, design a distributed estimation scheme to realize the orientation alignment and the absolute position estimation of free node $i$ merely using the local bearing measurement is the objective of this work.

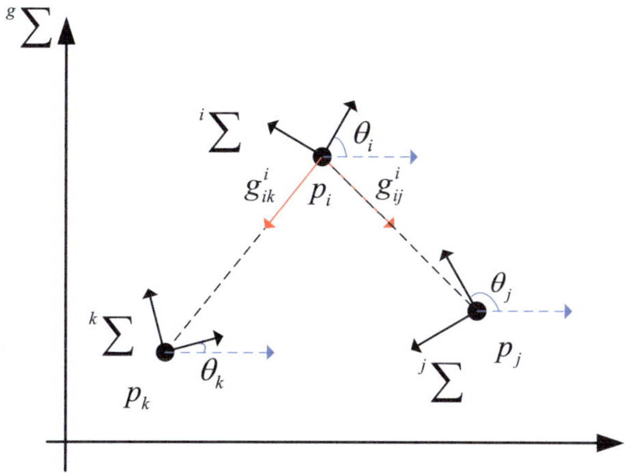


**Fig. 7.1** An example of the local bearing measurements

## 7.3 Main Results

### *7.3.1 Continuous-Time Estimator Design*

In this work, the absolute position information $p_i$ and the relative position information $p_i - p_j$ under the global reference frame ${}^{g}\sum$ cannot be measured, and the orientation angle $\theta_i \in [0, 2\pi]$ is also unknown. Therefore, denote $\hat{p}_i$ and $\hat{R}(\theta_i)$ are the estimates of absolute position $p_i$ and orientation matrix $R(\theta_i)$.

In order to realize the orientation alignment and the absolute position estimation of free node $i$, the distributed orientation estimation algorithm and localization estimation algorithm are proposed, respectively. The specific is shown below

$$\dot{\hat{g}}_{jk} = -Q_{\hat{g}_{jk}} \hat{R}(\theta_i) g^i_{jk}, \tag{7.1}$$

$$\dot{\hat{p}}_i = -Q_{\hat{g}_{ij}} \left(\hat{p}_i - \hat{p}_j\right) - Q_{\hat{g}_{ik}} \left(\hat{p}_i - \hat{p}_k\right), \tag{7.2}$$

where $j$ and $k$ are neighbors of $i$, $i > j$ and $i > k$. $\hat{g}_{jk} = \frac{\hat{p}_k - \hat{p}_j}{\|\hat{p}_k - \hat{p}_j\|}$, $\hat{p}_k$ and $\hat{p}_j$ represent the estimated position information of $j$ and $k$, respectively, and are received by $i$. $\hat{g}_{ij} = \hat{R}(\theta_i) g^i_{ij}$ is the estimated unit direction vector between nodes $i$ and $j$ in the global reference frame. $\hat{g}_{ik} = \hat{R}(\theta_i) g^i_{ik}$ is the estimated unit direction vector between nodes $i$ and $k$. $g^i_{ij}$ and $g^i_{ik}$, respectively, represent the unit direction vectors of $j$ and $k$ under the local reference frame ${}^{i}\sum$, which can be directly calculated based on the local bearing measurement.

In estimation schemes (7.1) and (7.2), all the available bearing information of the free node $i$ is measured based on its own local reference frame, and the free node $i$ only performs one-way local communication with neighboring nodes, so the proposed estimation algorithms (7.1) and (7.2) are also distributed.

To better explain the orientation estimation algorithm (7.1), a lemma is provided to show that the orientation angle $\theta_i$ can be estimated indirectly using the estimated unit direction vector $\hat{g}_{jk}$.

**Lemma 7.1** *Based on local bearing measurement $g^i_{jk}$, the proposed orientation estimation algorithm (7.1) can realize indirect estimation of the orientation angle $\theta_i$ through the estimation of unit direction vector $\hat{g}_{jk}$.*

***Proof*** In this work, the local bearing information can be directly measured. Since $\hat{g}_{jk} = \hat{R}(\theta_i)\, g^i_{jk}$, one can obtain that $\dot{\hat{g}}_{jk} = \dot{\hat{R}}(\theta_i)\, g^i_{jk}$.

Define $\dot{\hat{g}}_{jk} = \left(\dot{\hat{x}}, \dot{\hat{y}}\right)^{\mathrm{T}}$, $\hat{g}_{jk} = \left(\hat{x}, \hat{y}\right)^{\mathrm{T}}$ and $g^i_{jk} = (x_i, y_i)^{\mathrm{T}}$, then one has $\hat{R}(\theta_i) = \begin{bmatrix} x_i\hat{x} + y_i\hat{y} & x_i\hat{y} - \hat{x}y_i \\ \hat{x}y_i - x_i\hat{y} & x_i\hat{x} + y_i\hat{y} \end{bmatrix}$, $\dot{\hat{R}}(\theta_i) = \begin{bmatrix} x_i\dot{\hat{x}} + y_i\dot{\hat{y}} & x_i\dot{\hat{y}} - \dot{\hat{x}}y_i \\ \dot{\hat{x}}y_i - x_i\dot{\hat{y}} & x_i\dot{\hat{x}} + y_i\dot{\hat{y}} \end{bmatrix}$.

According to the above analysis, one can obtain that the orientation angle $\theta_i$ can be estimated indirectly by the estimated $\hat{g}_{jk}$.

The proof of Lemma 7.1 is completed.

**Remark 7.2** The orientation estimation algorithm (7.1) is an indirect method for estimating the orientation angle $\theta_i$. The estimated $\hat{g}_{jk}$ is used to estimate the rotation matrix $\hat{R}(\theta_i)$, then the orientation angle $\theta_i$ can be calculated according to the obtained rotation matrix. Eventually, $\hat{R}(\theta_i)$ will converge to $R(\theta_i)$ when $\hat{g}_{jk}$ converges to the true value $g_{jk}$.

### 7.3.2 Convergence Analysis

The convergence of the estimation algorithms (7.1)–(7.2) merely using the local bearing measurement is provided.

**Theorem 7.1** *If the sensor network satisfies Assumptions 7.1 and 7.2, the proposed distributed estimation algorithms (7.1) and (7.2) can guarantee that the estimated rotation matrix $\hat{R}(\theta_i)$ converges to the real rotation matrix $R(\theta_i)$ and the estimated position $\hat{p}_i$ converges to the absolute position $p_i$.*

Since the sensor network satisfies the Henneberg construction, the free node $i \in \{3, \ldots, n\}$ has two neighbors $j$ and $k$, $1 \le j \ne k \le i - 1$.

The brief proof ideas are given as follows:

At first, we will prove that $\hat{R}(\theta_3)$ converges to $R(\theta_3)$ and $\hat{p}_3$ converges to $p_3$. Note that the two neighbors of free node 3 are anchor nodes 1 and 2. Then, we will prove that $\hat{R}(\theta_i)$ converges to $R(\theta_i)$ and $\hat{p}_i$ converges to $p_i$.

***Proof*** The complete proof process of orientation estimation and position estimation of free node 3 are given as follows.

**Step 1** The convergence of estimated rotation matrix $\hat{R}(\theta_3)$.

For free node 3, the equilibrium point of the estimation algorithm (7.1) satisfies the following condition

$$0 = -Q_{g_{12}} \hat{R}(\theta_3) g_{12}^3. \tag{7.3}$$

Anchor nodes 1 and 2 know their absolute positions in ${}^g\sum$, so the unit vector $g_{12}$ can be directly calculated without estimation.

By left multiplying the matrix $(\hat{g}_{12} - g_{12})^{\mathrm{T}}$ on both sides of (7.3), one can obtain that

$$\begin{aligned} 0 &= -(\hat{g}_{12} - g_{12})^{\mathrm{T}} Q_{g_{12}} \hat{R}(\theta_3) g_{12}^3 \\ &= -\left(\hat{R}(\theta_3) g_{12}^3 - g_{12}\right)^{\mathrm{T}} Q_{g_{12}} \left(\hat{R}(\theta_3) g_{12}^3 - g_{12} + g_{12}\right). \end{aligned} \tag{7.4}$$

Since $Q_{g_{12}} g_{12} = 0$, (7.4) becomes

$$0 = -\left(\hat{R}(\theta_3) g_{12}^3 - g_{12}\right)^{\mathrm{T}} Q_{g_{12}} \left(\hat{R}(\theta_3) g_{12}^3 - g_{12}\right). \tag{7.5}$$

Equation (7.5) holds if and only if $\hat{R}(\theta_3) g_{12}^3 - g_{12} = 0$. In other words, the estimation algorithm (7.1) has a unique equilibrium point: $\hat{g}_{12} = g_{12}$.

Due to the fact that $\hat{g}_{12} = \hat{R}(\theta_3) g_{12}$ and $g_{12} = R(\theta_3) g_{12}$, one has

$$\left(\hat{R}(\theta_3) - R(\theta_3)\right) g_{12} = 0. \tag{7.6}$$

Denote $g_{12} = (x, y)^{\mathrm{T}}$, $\hat{R}(\theta_3) = \begin{bmatrix} \cos\hat{\theta}_3 & -\sin\hat{\theta}_3 \\ \sin\hat{\theta}_3 & \cos\hat{\theta}_3 \end{bmatrix}$ and $R(\theta_3) = \begin{bmatrix} \cos\theta_3 & -\sin\theta_3 \\ \sin\theta_3 & \cos\theta_3 \end{bmatrix}$, then (7.6) becomes

$$\begin{bmatrix} -2\sin\frac{\hat{\theta}_3+\theta_3}{2}\sin\frac{\hat{\theta}_3-\theta_3}{2} & -2\cos\frac{\hat{\theta}_3+\theta_3}{2}\sin\frac{\hat{\theta}_3-\theta_3}{2} \\ 2\cos\frac{\hat{\theta}_3+\theta_3}{2}\sin\frac{\hat{\theta}_3-\theta_3}{2} & -2\sin\frac{\hat{\theta}_3+\theta_3}{2}\sin\frac{\hat{\theta}_3-\theta_3}{2} \end{bmatrix} (x, y)^{\mathrm{T}} = 0. \tag{7.7}$$

Since $x$ and $y$ are not equal to zero at the same time, one has $\sin\frac{\hat{\theta}_3+\theta_3}{2}\sin\frac{\hat{\theta}_3-\theta_3}{2} = 0$ and $\cos\frac{\hat{\theta}_3+\theta_3}{2}\sin\frac{\hat{\theta}_3-\theta_3}{2} = 0$. Due to the fact that $\theta_3, \hat{\theta}_3 \in [0, 2\pi]$, a necessary and sufficient condition for the establishment of (7.7) is $\hat{\theta}_3 = \theta_3$, that is, $\hat{R}(\theta_3) = R(\theta_3)$.

Based on the above analysis (7.4)–(7.7), one can conclude that the estimation algorithm (7.1) has a unique equilibrium point, and at the equilibrium point, $\hat{R}(\theta_3) = R(\theta_3)$.

Consider the Lyapunov function

$$V = \frac{1}{2} \left\| \hat{g}_{12} - g_{12} \right\|^2. \tag{7.8}$$

Let $\lambda_{Q_{g_{12}}}$ be the minimum eigenvalue of $Q_{g_{12}}$. Since $Q_{g_{12}}$ is a positive semidefinite matrix, $\lambda_{Q_{g_{12}}} \geq 0$. Then, one has

$$
\begin{aligned}
\dot{V} &= -\left(\hat{R}(\theta_3)\, g_{12}^3 - g_{12}\right)^{\mathrm{T}} Q_{g_{12}} \left(\hat{R}(\theta_3)\, g_{12}^3 - g_{12}\right) \\
&\leq -\lambda_{Q_{g_{12}}} \left\| \hat{g}_{12} - g_{12} \right\|^2 \\
&= -\, 2\lambda_{Q_{g_{12}}} V.
\end{aligned} \tag{7.9}
$$

It follows from (7.9) that the estimation algorithm (7.1) converges exponentially at the unique equilibrium point, and $\hat{R}(\theta_3)$ exponentially converge to $R(\theta_3)$.

**Step 2** The convergence of estimated position $\hat{p}_3$.

For free node 3, the equilibrium point of the position estimation algorithm (7.2) satisfies

$$
0 = -Q_{\hat{g}_{31}} \left(\hat{p}_3 - p_1\right) - Q_{\hat{g}_{32}} \left(\hat{p}_3 - p_2\right). \tag{7.10}
$$

By left multiplying the matrix $\left(\hat{p}_3 - p_3\right)^{\mathrm{T}}$ on both sides of Eq. (7.10), one has

$$
\begin{aligned}
0 &= -\left(\hat{p}_3 - p_3\right)^{\mathrm{T}} Q_{\hat{g}_{31}} \left(\hat{p}_3 - p_1\right) - \left(\hat{p}_3 - p_3\right)^{\mathrm{T}} Q_{\hat{g}_{32}} \left(\hat{p}_3 - p_2\right) \\
&= -\left(\hat{p}_3 - p_3\right)^{\mathrm{T}} Q_{\hat{g}_{31}} \left(\hat{p}_3 - p_3\right) - \left(\hat{p}_3 - p_3\right)^{\mathrm{T}} Q_{\hat{g}_{32}} \left(\hat{p}_3 - p_3\right) \\
&+ \left(\hat{p}_3 - p_3\right)^{\mathrm{T}} Q_{\hat{g}_{31}} \left(p_1 - p_3\right) + \left(\hat{p}_3 - p_3\right)^{\mathrm{T}} Q_{\hat{g}_{32}} \left(p_2 - p_3\right).
\end{aligned} \tag{7.11}
$$

According to Step 1, one can know that $\hat{g}_{31}$ converges to $g_{31}$ and $\hat{g}_{32}$ converges to $g_{32}$ after applying the estimation algorithm (7.1). Since $Q_{g_{ij}} g_{ij} = 0$, one can obtain that $Q_{\hat{g}_{31}} \left(p_1 - p_3\right) = 0$ and $Q_{\hat{g}_{32}} \left(p_2 - p_3\right) = 0$. Then, (7.11) is equivalent to

$$
\begin{aligned}
0 &= -\left(\hat{p}_3 - p_3\right)^{\mathrm{T}} Q_{g_{31}} \left(\hat{p}_3 - p_3\right) - \left(\hat{p}_3 - p_3\right)^{\mathrm{T}} Q_{g_{32}} \left(\hat{p}_3 - p_3\right) \\
&= -\left(\hat{p}_3 - p_3\right)^{\mathrm{T}} \left(Q_{g_{31}} + Q_{g_{32}}\right) \left(\hat{p}_3 - p_3\right).
\end{aligned} \tag{7.12}
$$

Equation (7.12) holds if and only if $\hat{p}_3 - p_3 = 0$. Therefore, $\hat{p}_3 = p_3$ is the unique equilibrium point of the position estimation algorithm (7.2).

Consider the following Lyapunov function

$$
V = \frac{1}{2} \left\| \hat{p}_3 - p_3 \right\|^2. \tag{7.13}
$$

According to (7.11) and (7.12), the derivative of (7.13) is given as follows

$$
\begin{aligned}
\dot{V} &= -\left(\hat{p}_3 - p_3\right)^{\mathrm{T}} \left(Q_{g_{31}} + Q_{g_{32}}\right) \left(\hat{p}_3 - p_3\right) \\
&\leq -\lambda_{Q_{g_{31}} + Q_{g_{32}}} \left\| \hat{p}_3 - p_3 \right\|^2 \\
&= -2\lambda_{Q_{g_{31}} + Q_{g_{32}}} V,
\end{aligned} \tag{7.14}
$$

where $\lambda_{Q_{g_{31}} + Q_{g_{32}}}$ is the minimum eigenvalue of $(Q_{g_{31}} + Q_{g_{32}})$. Since $Q_{g_{31}}$ and $Q_{g_{32}}$ are positive semidefinite matrices, $\lambda_{Q_{g_{31}} + Q_{g_{32}}} \geq 0$.

From (7.14), one can conclude that the position estimation algorithm (7.2) converges exponentially at the unique equilibrium point, and $\hat{p}_3$ exponentially converge to $p_3$.

The complete proof process of the orientation estimation and the absolute position estimation of free node $i$ are shown below.

**Step 3** The convergence of estimated rotation matrix $\hat{R}(\theta_i)$.

The equilibrium point of (7.1) satisfies

$$0 = -P_{\hat{g}_{jk}} \hat{R}(\theta_i) g^{i}_{jk}. \tag{7.15}$$

By left multiplying the matrix $\left(\hat{g}_{jk} - g_{jk}\right)^{\mathrm{T}}$ on both sides of (7.15), one has

$$\begin{aligned} 0 &= -\left(\hat{g}_{jk} - g_{jk}\right)^{\mathrm{T}} Q_{\hat{g}_{jk}} \hat{R}(\theta_i) g^{i}_{jk} \\ &= -\left(\hat{R}(\theta_i) g^{i}_{jk} - g_{jk}\right)^{\mathrm{T}} Q_{\hat{g}_{jk}} \left(\hat{R}(\theta_i) g^{i}_{jk} - g_{jk} + g_{jk}\right). \end{aligned} \tag{7.16}$$

Based on the analysis (7.4) and (7.5), one can know that $\hat{g}_{jk}$ can converge to $g_{jk}$ when estimating the orientation angle of $i$, and due to the fact that $Q_{g_{jk}} g_{jk} = 0$, (7.16) becomes

$$0 = -\left(\hat{R}(\theta_i) g^{i}_{jk} - g_{jk}\right)^{\mathrm{T}} Q_{\hat{g}_{jk}} \left(\hat{R}(\theta_i) g^{i}_{jk} - g_{jk}\right). \tag{7.17}$$

It follows from (7.17) that orientation estimation algorithm (7.1) has a unique equilibrium point $\hat{g}_{jk} = g_{jk}$, and $\hat{R}(\theta_i) = R(\theta_i)$ at this equilibrium point.

Select a Lyapunov function

$$V = \frac{1}{2} \left\| \hat{g}_{jk} - g_{jk} \right\|^2. \tag{7.18}$$

The derivative of (7.18) is shown as

$$\begin{aligned} \dot{V} &= -\left(\hat{R}(\theta_i) g^{i}_{jk} - g_{jk}\right)^{\mathrm{T}} Q_{\hat{g}_{jk}} \left(\hat{R}(\theta_i) g^{i}_{jk} - g_{jk}\right) \\ &\leq -\lambda_{Q_{\hat{g}_{jk}}} \left\| \hat{g}_{jk} - g_{jk} \right\|^2 \\ &= -2\lambda_{Q_{\hat{g}_{jk}}} V, \end{aligned} \tag{7.19}$$

where $\lambda_{Q_{\hat{g}_{jk}}}$ represents the minimum eigenvalue of $Q_{\hat{g}_{jk}}$, and $\lambda_{Q_{\hat{g}_{jk}}} \geq 0$.

From (7.19), one can conclude that (7.1) converges exponentially at the unique equilibrium point $\hat{g}_{jk} = g_{jk}$, and $\hat{R}(\theta_i)$ exponentially converge to $R(\theta_i)$ eventually.

**Step 4** The convergence of estimated position $\hat{p}_i$.

The equilibrium point of (7.2) satisfies

$$0 = -Q_{\hat{g}_{ij}} \left(\hat{p}_i - \hat{p}_j\right) - Q_{\hat{g}_{ik}} \left(\hat{p}_i - \hat{p}_k\right). \tag{7.20}$$

By left multiplying the matrix $\left(\hat{p}_i - p_i\right)^{\mathrm{T}}$ on both sides of (7.20), one can obtain that

$$\begin{aligned} 0 &= -\left(\hat{p}_i - p_i\right)^{\mathrm{T}} Q_{\hat{g}_{ij}} \left(\hat{p}_i - \hat{p}_j\right) - \left(\hat{p}_i - p_i\right)^{\mathrm{T}} Q_{\hat{g}_{ik}} \left(\hat{p}_i - \hat{p}_k\right) \\ &= -\left(\hat{p}_i - p_i\right)^{\mathrm{T}} Q_{\hat{g}_{ij}} \left(\hat{p}_i - p_i\right) - \left(\hat{p}_i - p_i\right)^{\mathrm{T}} Q_{\hat{g}_{ik}} \left(\hat{p}_i - p_i\right) \\ &\quad + \left(\hat{p}_i - p_i\right)^{\mathrm{T}} Q_{\hat{g}_{ij}} \left(\hat{p}_j - p_i\right) + \left(\hat{p}_i - p_i\right)^{\mathrm{T}} Q_{\hat{g}_{ik}} \left(\hat{p}_k - p_i\right). \end{aligned} \tag{7.21}$$

In this work, the considered sensor network satisfies the Henneberg construction and nodes 1 and 2 are the anchor nodes, so the estimated orientation and position of free node 3 will converge first, and then gradually expand outward. Therefore, the estimated information $\hat{g}_{ij}$, $\hat{g}_{ik}$, $\hat{p}_j$ and $\hat{p}_k$ has converged to their true value when estimating the position of free node $i$. Based on the property of orthogonal projection matrix $Q_{g_{ij}} g_{ij} = 0$, one has $Q_{\hat{g}_{ij}} \left(\hat{p}_j - p_i\right) = 0$ and $Q_{\hat{g}_{ik}} \left(\hat{p}_k - p_i\right) = 0$.

According to the above analysis, (7.21) is equivalent to

$$\begin{aligned} 0 &= -\left(\hat{p}_i - p_i\right)^{\mathrm{T}} Q_{g_{ij}} \left(\hat{p}_i - p_i\right) - \left(\hat{p}_i - p_i\right)^{\mathrm{T}} Q_{g_{ik}} \left(\hat{p}_i - p_i\right) \\ &= -\left(\hat{p}_i - p_i\right)^{\mathrm{T}} \left(Q_{g_{ij}} + Q_{g_{ik}}\right) \left(\hat{p}_i - p_i\right). \end{aligned} \tag{7.22}$$

From (7.22), one can know that the unique equilibrium point of (7.2) is $\hat{p}_i = p_i$. Design a Lyapunov function,

$$V = \frac{1}{2} \left\| \hat{p}_i - p_i \right\|^2. \tag{7.23}$$

Based on the analysis from (7.21) to (7.22), the derivative of (7.23) is given as follows

$$\begin{aligned} \dot{V} &= -\left(\hat{p}_i - p_i\right)^{\mathrm{T}} \left(Q_{\hat{g}_{ij}} + Q_{\hat{g}_{ik}}\right) \left(\hat{p}_i - p_i\right) \\ &\leq -\lambda_{Q_{\hat{g}_{ij}} + Q_{\hat{g}_{ik}}} \left\| Q_{\hat{g}_{ij}} + Q_{\hat{g}_{ik}} \right\|^2 \\ &= -2\lambda_{Q_{\hat{g}_{ij}} + Q_{\hat{g}_{ik}}} V, \end{aligned} \tag{7.24}$$

where $\lambda_{Q_{\hat{g}_{ij}} + Q_{\hat{g}_{ik}}} \geq 0$ represents the minimum eigenvalue of the matrix $Q = Q_{\hat{g}_{ij}} + Q_{\hat{g}_{ik}}$. One can obtain that (2) converges exponentially at the equilibrium point $\hat{p}_i = p_i$, and $\hat{p}_i$ exponentially converges to $p_i$ eventually.

The proof of Theorem 7.1 is completed.

**Remark 7.3** In the proof of Theorem 7.1, the sensor network is regarded as a cascade system. The convergence of the orientation angle estimation information $\theta_i$ depends on the position information $\hat{p}_j$ and $\hat{p}_k$ of the neighbor nodes $j$ and $k$, $1 \leq j \neq k \leq i - 1$. Meanwhile, the convergence of the position estimation information of $i$ depends on the accurate orientation angle estimation information $R\left(\theta_i\right)$. The proof of convergence starts from free node 3, and the neighbors of node 3 are anchor nodes 1 and 2 whose absolute positions are known. Therefore, the convergence of the orientation angle and the absolute position of the free nodes in the cascade system can be proved progressively using the Mathematical induction.

## 7.4 Numerical Simulation

In this section, two simulation results are provided to illustrate the effectiveness of the proposed algorithms. The orientations of all free nodes are given randomly, and the absolute positions of all nodes satisfy the Henneberg construction.

In the simulation, only the local bearing measurements is used to align the orientation and estimate the absolute position. Compared with the distributed algorithm based on local bearing and relative orientation measurements, the relative orientation measurement is not needed. And it may be difficult to directly measure relative orientations in practice in the case of sensor node has no knowledge of global reference frame.

First, we consider a 10 nodes sensor network with directed acyclic graph generated by Henneberg construction. As shown in Fig. 7.2, the red circles are the anchor nodes, and the red-red lines represent the global reference frame of the anchor node. The blue circles are the free nodes, and the black-black lines indicate the local reference frames of the free node.

Applying the proposed estimation algorithms (7.1) and (7.2), the initial estimated values are given randomly, and the simulation results are shown in Figs. 7.2, 7.3

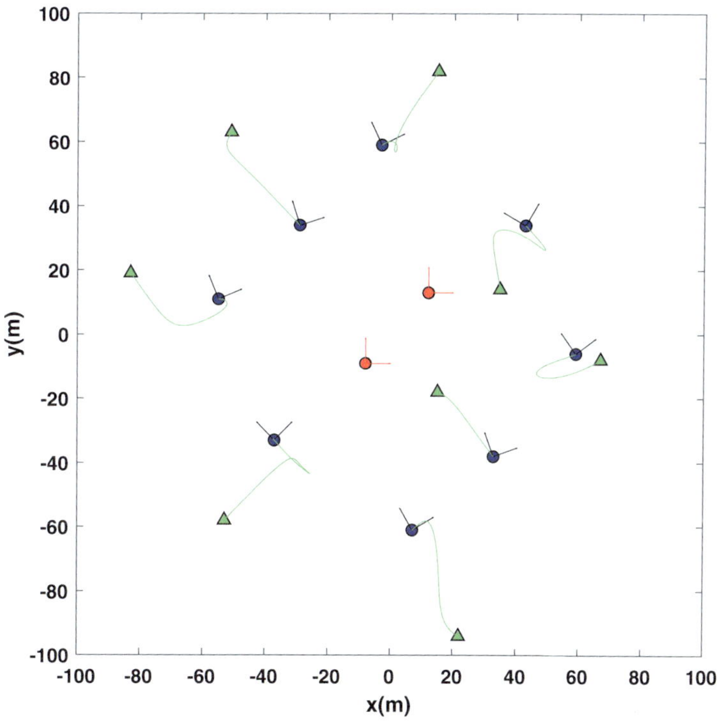


**Fig. 7.2** The convergence trajectories of the estimated positions of eight free nodes

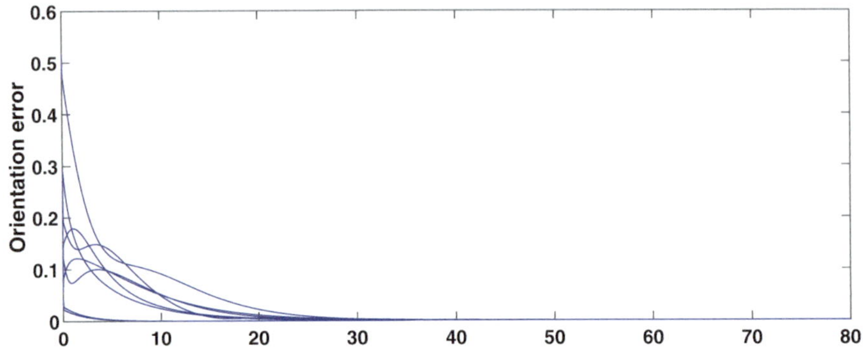


**Fig. 7.3** The orientation estimation error curves of eight free nodes

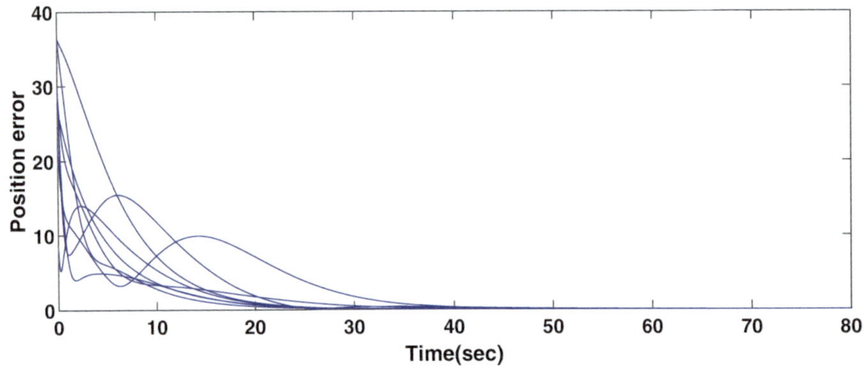


**Fig. 7.4** The position estimation error curves of eight free nodes

and 7.4. The convergence trajectories of the free nodes are shown in Fig. 7.2, where the green triangles are the initial estimated positions, the green curves represent the convergence process of the estimated positions. As can be seen in Fig. 7.3, the orientation estimation errors $\left\| \hat{R}(\theta_i) - R(\theta_i) \right\|$ of free nodes converge to zero eventually. The position estimation errors $\| \hat{p}_i - p_i \|$ are shown in Fig. 7.4, it can be found that the free nodes can estimate their absolute position in the global reference frame. From Figs. 7.2, 7.3 and 7.4, one can conclude that the proposed estimation algorithms (7.1) and (7.2) can guarantee the accurate orientation and position estimation merely using the local bearing measurement.

Next, we consider a new 18 nodes sensor network with 2 anchor nodes and 16 free nodes. The new sensor network generated by adding new vertices and splitting edges, which is satisfies the Henneberg construction. Applying estimation algorithms (7.1) and (7.2), the simulation results are shown in Figs. 7.5, 7.6 and 7.7. It can be found that although the convergence speed slows down with the increase of the scale of

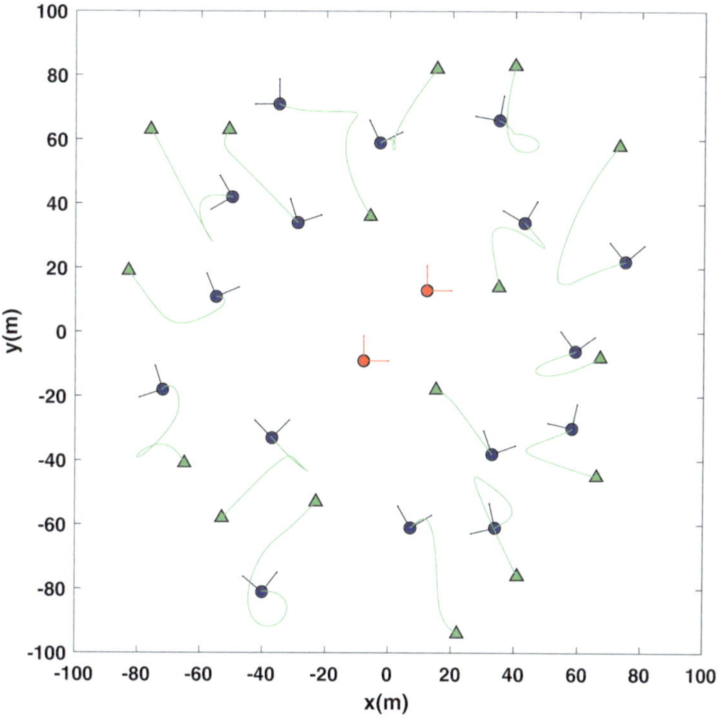


**Fig. 7.5** The convergence trajectories of the estimated positions of sixteen free nodes

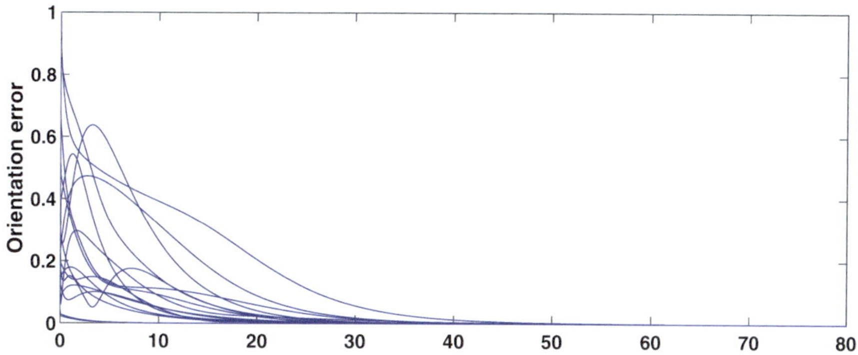


**Fig. 7.6** The orientation estimation error curves of sixteen free nodes

sensor networks, both the orientation and the position estimation errors converge to zero eventually.

Through the simulation results, one can draw the conclusions that: (a) the proposed estimation algorithms can realize the accurate orientation estimation and the absolute

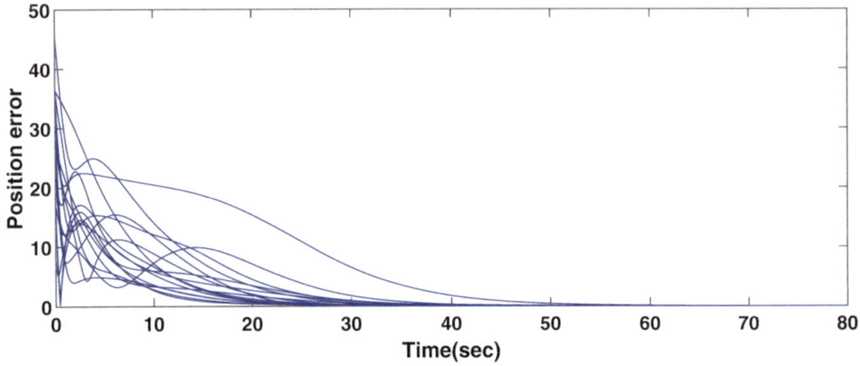


**Fig. 7.7** The position estimation error curves of sixteen free nodes

position localization of the sensor networks. (b) the proposed estimation algorithms are also applicable to large-scale sensor networks.

## 7.5 Conclusions

In this chapter, a new continuous-time distributed estimation scheme is proposed to solve the sensor network localization problem merely using the local bearing measurement. Each sensor node can estimate its accurate orientation and absolute position based on the local bearing measurement and the communication with the neighbors, without the knowledge of the global reference frame. Based on cascade system, the global convergence of the sensor network is proved by introducing Mathematical Induction. The simulation results also prove that the proposed estimation algorithms are suitable for large-scale sensor networks. Several interesting problems in local-bearing-only based should further investigation. For example, the formation control of multi-agent systems, the distributed localization estimation under time-varying communication networks, and the localization estimation in three-dimensional space in the noisy environment.

# Chapter 8
# Local-Bearing-Based Prescribed-Time Localization of MASs with Noisy Measurement

This chapter investigates stability and localizability of local-bearing-based multi-agent systems without common orientation in the presence of noisy measurement, which are more general but also more challenging to deal with than global-bearing-based multi-agent systems under ideal environment. Based on local-bearing unbiased estimator constructed from the historical information and a newly designed time-varying gain, a new robust prescribed-time orientation estimation algorithm is proposed to ensure that the local reference frame of the follower agent is aligned with the global one. The local bearing information is more easily obtained than global one. Therefore, the new orientation estimation result is expected to be more widely applicable. The robust orientation estimation algorithm is then applied to the problem of localization estimation, and a prescribed-time distributed localization estimation algorithm is developed. The distinctive advantage of this work is that only the local bearing information is used, and the fast and controllable localization estimation is achieved. The global convergence is derived based on the cascade system and some simulation results are provided to prove the effectiveness of the proposed estimation algorithms.

## 8.1 Introduction

The localization estimation based on bearing information is investigated and the necessary and sufficient conditions for network localizability are derived by combining the algebraic theory and rigid theory. One of the technical difficulties of localization estimation is that each agent has its own local reference frame and is not aligned with the global one. Therefore, it is of great significance to study the localization estimation of the agent in the case of the global reference frame unknown.

The orientation estimation problem has also been discussed in recent years. In [1], the inter-neighbor relative orientation information is used to align the orientations,

 
Y. Lv et al., *Distributed Localization and Formation Control of Multi-Agent Systems with Local Sensing Information*, https://doi.org/10.1007/978-981-96-6015-5_8

which require all the initial orientation angles must be less than $\pi$. In [2], the initial orientation angles constraint is removed, the algorithm requires a Gram-Schmidt orthogonalization at each time instance. In a word, the agent aligns the orientation based on bearing information, then the estimated absolute orientations can be used to solve the formation control and localization estimation problems.

It is worth noting that all the above works are asymptotically or exponentially stables, which means that the upper bounded of the estimation convergence time is not guaranteed. In practical applications, as a link between the preceding and the following in the whole task, the speed of localization will directly affect the efficiency of the whole MASs, which should be taken into account. With the proposal of homogeneous theory and Lyapunov Stability criterion, finite time consistency theory has been developed rapidly. It should be pointed out that the upper bound of the convergence time of the finite time consistency algorithm is related to the initial state of each agent. In order to eliminate the influence of initial state on the convergence time, fixed time consistency algorithm has been studied [5, 6]. Further, in order to directly predict and set the specific convergence time of the system state, in recent years, the specified time consistency algorithm has received more attention [7, 8]. Note that, these works are focused on the design of control strategy, and few results on localization estimation.

In the aforementioned distributed localization estimation, the environment is assumed to be ideal. However, in real applications, measurement noise is ubiquitous. Low sensor accuracy, sudden changes in application environment, and other factors may cause measurement noise. The localizability and stability problem of networked systems with measurement noise have been investigated until fairly recently.

To be specific, the stability and localizability of local-bearing-based MASs under measurement noise is investigated in this work. The contributions can be summarized as follows:

(1) A robust prescribed-time orientation estimation algorithm is proposed, which guarantees the orientation alignment under measurement noise.
(2) A prescribed-time distributed localization estimation algorithm merely using local bearing information is designed, which realizes fast and controllable position estimation.
(3) The constraint of initial orientation angles is removed, the Gram-Schmidt orthogonalization is not required, and the relative orientation measurements are not need.

The rest of this chapter is organized as follows. Section 8.2 shows the problem formulation. In Sect. 8.3 gives the main results. Simulation and experiment results are presented in Sect. 8.4 and Conclusions are given in Sect. 8.5.

## 8.2 Problem Formulation

Consider a $n$ agents multi-agent systems in two dimensional plane, where two agents are leaders (labelled as 1 and 2) and the remaining $(n-2)$ are follower agents (labelled as $\{3, \ldots, n\}$). The global absolute position of leader agent in ${}^{g}\Sigma$ can be directly obtained, while the follower agent needs to be estimated.

In this work, the follower agent has its own local reference frame, which is not aligned with ${}^{g}\Sigma$, as shown in Fig. 8.1. Each follower agent can measure the bearing information of its neighbors (we call it local bearing information) and communicate with its neighbors, i.e., receive the position information sent by the neighbors. As mentioned in introduction part, the measured local bearing information contains noise due to the complex application environment and low sensor accuracy. If the measurement noise is not properly handled, as minor as, the final localization accuracy will decline, or it may even cause the task failure performed by the MASs.

The objective of this work is to design a robust distributed localization estimation scheme to realize the orientation alignment and the localization estimation of MASs under bounded measurement noise, i.e., $\lim_{t\to T} \hat{R}(\theta_i) = R(\theta_i)$ and $\lim_{t\to T} \hat{c}_i = c_i$, where $T$ is the prescribed convergence time, $\hat{R}(\theta_i)$ is the estimated rotation matrix and $\hat{c}_i$ denotes the estimated position information of agent $i$.

## 8.3 Prescribed-Time Distributed Localization Estimation

In this section, the localization estimation of MASs under bounded measurement noise is considered. The local bearing information with measurement noise can be expressed as

$$\tilde{b}^{i}_{jk}(t) = b^{i}_{jk}(t) + \sigma^{i}_{jk}(t), \tag{8.1}$$

where $b^{i}_{jk}(t)$ denotes the real local bearing information, $\sigma^{i}_{jk}(t)$ represents the measurement noise, $t$ is the sampling time.

Two assumptions are given as follows.

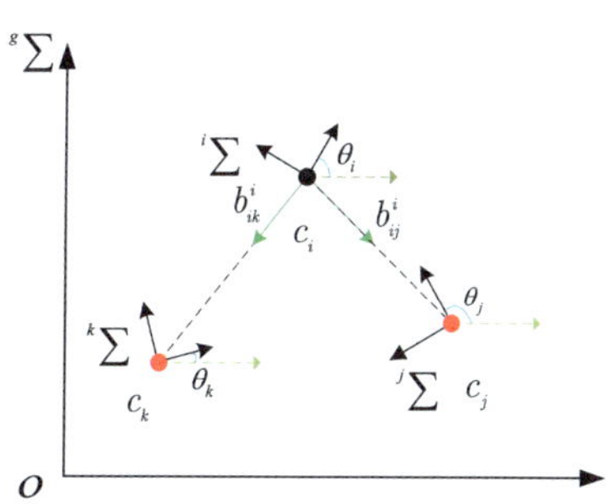


**Fig. 8.1** The local bearing measurements

**Assumption 8.1** The directed topology graph $\mathcal{G} = (\nu, \varepsilon)$ of the MASs satisfies the Henneberg construction.

**Remark 8.1** Henneberg construction meets the conditions of bearing localizability [3] and the self-localizability [4] of MASs. In addition, Henneberg construction can be infinitely split in theory, which is suitable for large-scale networked systems.

**Assumption 8.2** For all agent $i, j, k \in v$, the measurement noise $\sigma^i_{jk}(t)$ is bounded and independent identically distributed, $E\left(\sigma^i_{jk}(t)\right) = 0$, $E$ represents the mathematical expectation.

Before the estimator design, the measured local bearing information $\tilde{b}^i_{jk}(t)$ needs to be processed first.

$$\hat{b}^i_{jk}(t) = \frac{1}{t}\sum_{\tau=0}^{t} \tilde{b}^i_{jk}(\tau), \hat{b}^i_{jk}(1) = \tilde{b}^i_{jk}(1). \tag{8.2}$$

After applying (8.2) and according to strong law of large numbers, one has $\lim\limits_{t\to\infty} \hat{b}^i_{jk}(t) = b^i_{jk}(t)$.

## 8.3.1 Prescribed-Time Estimator Design

A robust prescribed-time distributed orientation estimation algorithm and a prescribed-time distributed position estimation algorithm are proposed, respectively. The specific is shown below,

$$\dot{\hat{b}}_{jk} = -\gamma\left(\alpha + \beta\frac{\dot{\mu}}{\mu}\right)\left(Q_{\tilde{b}_{jk}} \hat{R}\left(\theta_i\right) \hat{b}^i_{jk}\right), \tag{8.3}$$

$$\dot{\hat{c}}_i = -\left(\alpha + \beta\frac{\dot{\mu}}{\mu}\right)\left(Q_{\hat{b}_{ij}}\left(\hat{c}_i - \hat{c}_j\right) + Q_{\hat{b}_{ik}}\left(\hat{c}_i - \hat{c}_k\right)\right), \tag{8.4}$$

where $\gamma(t) = \frac{1}{(1+t)^{\lambda}}$ is a time-varying factor, $0 < \lambda < \frac{1}{2}$. $\alpha, \beta > 0$, $\mu(t) = \begin{cases} \frac{T^h}{(T-t)^h}, t \in [0, T) \\ 1, t \in [T, \infty) \end{cases}$ and $\dot{\mu}(t) = \begin{cases} \frac{h}{T}\mu^{\left(1+\frac{1}{h}\right)}, t \in [0, T) \\ 0, t \in [T, \infty) \end{cases}$, $h > 0$, $T$ is the prescribed convergence time. $\tilde{b}_{jk} = \frac{\hat{c}_k - \hat{c}_j}{\|\hat{c}_k - \hat{c}_j\|}$, $\hat{c}_k$ and $\hat{c}_j$ are the estimated position information of agents $k$ and $j$. $\hat{b}_{ij} = \hat{R}\left(\theta_i\right)\hat{b}^i_{ij}$, $\hat{b}_{ik} = \hat{R}\left(\theta_i\right)\hat{b}^i_{ik}$.

**Remark 8.2** Based on local bearing measurement, the proposed robust prescribed-time distributed orientation estimation algorithm (8.3) can realize indirect estimation of the orientation angle $\theta_i$ through the estimation of unit direction vector $\hat{b}_{jk}$.

In this work, the local bearing information can be directly measured. Since $\hat{b}_{jk} = \hat{R}(\theta_i) b^i_{jk}$, one can obtain that $\dot{\hat{b}}_{jk} = \dot{\hat{R}}(\theta_i) b^i_{jk}$.

Define $\dot{\hat{b}}_{jk} = \left(\dot{\hat{\varphi}}_x, \dot{\hat{\varphi}}_y\right)^{\mathrm{T}}$, $\hat{b}_{jk} = \left(\hat{\varphi}_x, \hat{\varphi}_y\right)^{\mathrm{T}}$ and $b^i_{jk} = \left(\varphi^i_x, \varphi^i_y\right)^{\mathrm{T}}$, then one has $\hat{R}(\theta_i) = \begin{bmatrix} \varphi^i_x\hat{\varphi}_x + \varphi^i_y\hat{\varphi}_y & \hat{\varphi}_x\varphi^i_y - \varphi^i_x\hat{\varphi}_y \\ -\hat{\varphi}_x\varphi^i_y + \varphi^i_x\hat{\varphi}_y & \varphi^i_y\hat{\varphi}_y + \varphi^i_x\hat{\varphi}_x \end{bmatrix}$, $\dot{\hat{R}}(\theta_i) = \begin{bmatrix} \varphi^i_x\dot{\hat{\varphi}}_x + \varphi^i_y\dot{\hat{\varphi}}_y & \dot{\hat{\varphi}}_x\varphi^i_y - \varphi^i_x\dot{\hat{\varphi}}_y \\ -\dot{\hat{\varphi}}_x\varphi^i_y + \varphi^i_x\dot{\hat{\varphi}}_y & \varphi^i_y\dot{\hat{\varphi}}_y + \varphi^i_x\dot{\hat{\varphi}}_x \end{bmatrix}$.

Based on the above analysis, one can conclude that the orientation angle $\theta_i$ can be estimated indirectly by the estimated unit direction vector $\hat{b}_{jk}$.

### 8.3.2 Convergence Analysis

The convergence of the robust prescribed-time localization estimation scheme is shown below.

**Theorem 8.1** *Suppose the MASs satisfies Assumptions 8.1 and 8.2, applying the robust prescribed-time distributed estimation algorithms (8.3) and (8.4), one can guarantee that the estimated rotation matrix* $\hat{R}(\theta_i)$ *converges to the real rotation matrix* $R(\theta_i)$ *and the estimated position* $\hat{c}_i$ *converges to the global absolute position* $c_i$ *at the specified time.*

In this section, the mathematical induction is introduced to prove Theorem 8.1. Since the MASs satisfy the Henneberg construction, the follower agent $i \in \{3, \cdots, n\}$ has two neighbors $j$ and $k$, $1 \le j \ne k \le i-1$. Note that, the neighbors of follower agent 3 are leader agents 1 and 2.

***Proof*** The proof is mainly divided into two parts, the specific is given below.

In the first part, we will prove that $\hat{R}(\theta_3)$ converges to $R(\theta_3)$ and $\hat{c}_3$ converges to $c_3$ at the specified time. In the second part, we will prove that $\hat{R}(\theta_i)$ converges to $R(\theta_i)$ and $\hat{c}_i$ converges to $c_i$ at the specified time.

According to Assumption 8.2 and applying the unbiased estimator (8.2), it is reasonable to replace $\hat{b}^i_{jk}(t)$ with $\delta b^i_{jk}$, where $|\delta| > 0$.

**Part** I. The convergence of the estimated orientation rotation matrix $\hat{R}(\theta_3)$ and the estimated position $\hat{c}_3$.

For follower agent 3, the equilibrium points of (8.3) satisfies

$$\begin{aligned} 0 &= -\gamma\left(\alpha + \beta\frac{\dot{\mu}}{\mu}\right)\left(Q_{b_{12}}\hat{R}(\theta_3)\hat{b}^3_{12}\right) \\ &= -\gamma\delta\left(\alpha + \beta\frac{\dot{\mu}}{\mu}\right)\left(Q_{b_{12}}\hat{R}(\theta_3)b^3_{12}\right). \end{aligned} \tag{8.5}$$

Multiply $\left(\hat{b}_{12} - b_{12}\right)^{\mathrm{T}}$ on both sides of (8.5), one can obtain that

$$0 = -\gamma\delta\left(\alpha + \beta\frac{\dot{\mu}}{\mu}\right)\left(\left(\hat{b}_{12} - b_{12}\right)^{\mathrm{T}} Q_{b_{12}} \hat{R}\left(\theta_3\right) b_{12}^3\right)$$
$$= -\gamma\delta\left(\alpha + \beta\frac{\dot{\mu}}{\mu}\right)\left(\left(\hat{R}\left(\theta_3\right) b_{12}^3 - b_{12}\right)^{\mathrm{T}} Q_{b_{12}} \hat{R}\left(\theta_3\right) b_{12}^3\right). \tag{8.6}$$

Due to the fact that $Q_{b_{12}} b_{12} = 0$, (8.6) becomes

$$0 = -\gamma\delta\left(\alpha + \beta\frac{\dot{\mu}}{\mu}\right)\times$$
$$\left(\left(\hat{R}\left(\theta_3\right) b_{12}^3 - b_{12}\right)^{\mathrm{T}} Q_{b_{12}}\left(\hat{R}\left(\theta_3\right) b_{12}^3 - b_{12}\right)\right). \tag{8.7}$$

Since $\left(\alpha + \beta\frac{\dot{\mu}}{\mu}\right) > 0$, $\delta \neq 0$ and $\gamma > 0$, (8.7) holds if and only if $\hat{R}\left(\theta_3\right) b_{12}^3 - b_{12} = 0$. Substitute $b_{12}^3 = R(\theta_3)^{-1} b_{12}$ into $\hat{R}\left(\theta_3\right) b_{12}^3 - b_{12} = 0$, one has

$$\left(\hat{R}\left(\theta_3\right) - R\left(\theta_3\right)\right) b_{12} = 0. \tag{8.8}$$

Equation (8.8) means that $\hat{R}\left(\theta_3\right) = R\left(\theta_3\right)$ is the unique equilibrium point of (8.3). Select a Lyapunov function as follows

$$V = \frac{1}{2}\left\|\hat{b}_{12} - b_{12}\right\|^2. \tag{8.9}$$

Take the derivative of $V$ and based on the analysis of (8.7), one can obtain that

$$\dot{V} = -\gamma\delta\left(\alpha + \beta\frac{\dot{\mu}}{\mu}\right)\left(\left(\hat{R}\left(\theta_3\right) b_{12}^3 - b_{12}\right)^{\mathrm{T}} Q_{b_{12}}\left(\hat{R}\left(\theta_3\right) b_{12}^3 - b_{12}\right)\right)$$
$$\leq -\gamma\delta\left(\alpha + \beta\frac{\dot{\mu}}{\mu}\right)\lambda_{Q_{b_{12}}}\left\|\hat{b}_{12} - b_{12}\right\|^2$$
$$= -2\gamma\delta\left(\alpha + \beta\frac{\dot{\mu}}{\mu}\right)\lambda_{Q_{b_{12}}} V$$
$$= -2\gamma\delta\alpha\lambda_{Q_{b_{12}}} V - 2\gamma\delta\beta\lambda_{Q_{b_{12}}}\frac{\dot{\mu}}{\mu} V, \tag{8.10}$$

where $\lambda_{Q_{b_{12}}} \geq 0$ is the minimum eigenvalue of $Q_{b_{12}}$.

From (8.10), one can obtain that $\dot{V} \leq -\alpha_1 V - \beta_1\frac{\dot{\mu}}{\mu} V$, where $\alpha_1 = 2\gamma\delta\alpha\lambda_{Q_{b_{12}}}$ and $\beta_1 = 2\gamma\delta\beta\lambda_{Q_{b_{12}}}$. According to Lemma 1.1, one has $V\left(t\right)\begin{cases}\leq \mu^{-\beta_1} e^{-\alpha_1 t} V\left(0\right), t \in [0, T) \\ \equiv 0, t \in [T, \infty)\end{cases}$. Then, one can conclude that $\hat{b}_{12}$ can converge to $b_{12}$ at the specified time $T$, that is $\hat{R}\left(\theta_3\right)$ converges to $R\left(\theta_3\right)$.

Next, we will prove $\hat{c}_3$ converges to $c_3$.

The equilibrium points of (8.4) satisfies

$$0 = -\left(\alpha + \beta\frac{\dot{\mu}}{\mu}\right)\left(Q_{\hat{b}_{31}}\left(\hat{c}_3 - c_1\right) + Q_{\hat{b}_{32}}\left(\hat{c}_3 - c_2\right)\right). \tag{8.11}$$

Multiply $(\hat{c}_3 - c_3)^{\mathrm{T}}$ on both sides of (8.11), one has

$$\begin{aligned} 0 &= -\left(\alpha + \beta\frac{\dot{\mu}}{\mu}\right)(\hat{c}_3 - c_3)^{\mathrm{T}}\left(Q_{\hat{b}_{31}}(\hat{c}_3 - c_1) + Q_{\hat{b}_{32}}(\hat{c}_3 - c_2)\right) \\ &= -\left(\alpha + \beta\frac{\dot{\mu}}{\mu}\right)(\hat{c}_3 - c_3)^{\mathrm{T}} \times \\ &\left(Q_{\hat{b}_{31}}(\hat{c}_3 - c_3 + c_3 - c_1) + Q_{\hat{b}_{32}}(\hat{c}_3 - c_3 + c_3 - c_2)\right). \end{aligned} \tag{8.12}$$

According to the convergence analysis of $\hat{R}(\theta_3)$, one can know that $\hat{b}_{31} \to b_{31}$ and $\hat{b}_{32} \to b_{32}$ when $t \to T$. Then, (8.12) becomes

$$0 = -\left(\alpha + \beta\frac{\dot{\mu}}{\mu}\right)(\hat{c}_3 - c_3)^{\mathrm{T}}\left(Q_{b_{31}} + Q_{b_{32}}\right)(\hat{c}_3 - c_3). \tag{8.13}$$

Equation (8.13) holds if and only if $\hat{c}_3 = c_3$.

Select a Lyapunov function

$$V = \frac{1}{2}\left\|\hat{c}_3 - c_3\right\|^2. \tag{8.14}$$

Take the derivative of $V$, yields

$$\begin{aligned} \dot{V} &= -\left(\alpha + \beta\frac{\dot{\mu}}{\mu}\right)(\hat{c}_3 - c_3)^{\mathrm{T}}\left(Q_{b_{31}} + Q_{b_{32}}\right)(\hat{c}_3 - c_3) \\ &\leq -2\left(\alpha + \beta\frac{\dot{\mu}}{\mu}\right)\lambda_{Q_{b_{31}}+Q_{b_{32}}} V, \end{aligned} \tag{8.15}$$

where $\lambda_{Q_{b_{31}}+Q_{b_{32}}}$ is the minimum eigenvalue of $\left(Q_{b_{31}} + Q_{b_{32}}\right)$, $\lambda_{Q_{b_{31}}+Q_{b_{32}}} \geq 0$.

From (8.15), one has $\dot{V} \leq -\alpha_2 V - \beta_2\frac{\dot{\mu}}{\mu}V$, where $\alpha_2 = 2\alpha\lambda_{Q_{b_{31}}+Q_{b_{32}}}$ and $\beta_2 = 2\beta\lambda_{Q_{b_{31}}+Q_{b_{32}}}$.

According to Lemma 1.1, one can obtain that $V(t)\begin{cases} \leq \mu^{-\beta_2}e^{-\alpha_2 t}V(0), t \in [0, T) \\ \equiv 0, t \in [T, \infty) \end{cases}$.
Then, one can conclude that $\hat{c}_3$ can converge to $c_3$ at the specified time $2T$.

Based on the above analysis, one can conclude that the estimated rotation matrix $\hat{R}(\theta_3)$ and the estimated position $\hat{c}_3$ can converge to the real one at the specified time. The upper limit of specified convergence time is $2T$ in theory.

The proof of Part I is completed.

**Remark 8.3** In the proof of Part I, the analysis of the upper limit of convergence time is based on a cascaded system. For agent 3, the orientation estimation and position estimation are synchronous. Although the convergence time of $\hat{c}_3$ depends on the convergence of the estimated rotation matrix $\hat{R}(\theta_3)$, after orientation alignment, the estimated position $\hat{c}_3$ can converge to the real one in a short time interval. Therefore, in practical applications, the convergence time of agent 3 is less than $2T$ and close to $T$.

**Part** II. The convergence of $\hat{R}(\theta_i)$ and $\hat{c}_i$.

For follower agent $i$, the equilibrium points of (8.3) satisfies

$$0 = -\gamma\left(\alpha + \beta\frac{\dot{\mu}}{\mu}\right)\left(Q_{\tilde{b}_{jk}}\hat{R}(\theta_i)\,\delta b^i_{jk}\right). \tag{8.16}$$

Multiply $\left(\hat{b}_{jk} - b_{jk}\right)^{\mathrm{T}}$ on both sides of (8.16),

$$\begin{aligned}0 &= -\gamma\delta\left(\alpha + \beta\tfrac{\dot{\mu}}{\mu}\right)\left(\left(\hat{R}(\theta_i)\,b^i_{jk} - b_{jk}\right)^{\mathrm{T}} Q_{\tilde{b}_{jk}}\hat{R}(\theta_i)\,b^i_{jk}\right)\\ &= -\gamma\delta\left(\alpha + \beta\tfrac{\dot{\mu}}{\mu}\right)\times\\ &\left(\left(\hat{R}(\theta_i)\,b^i_{jk} - b_{jk}\right)^{\mathrm{T}} Q_{\tilde{b}_{jk}}\left(\hat{R}(\theta_i)\,b^i_{jk} - b_{jk} + b_{jk}\right)\right).\end{aligned} \tag{8.17}$$

Based on the analysis of Part I, one can obtain that $\lim_{t\to 2(i-3)T} \tilde{b}_{jk} = b_{jk}$, and due to the fact that $Q_{b_{jk}} b_{jk} = 0$, then (8.17) becomes

$$0 = -\gamma\delta\left(\alpha + \beta\tfrac{\dot{\mu}}{\mu}\right)\left(\left(\hat{R}(\theta_i)\,b^i_{jk} - b_{jk}\right)^{\mathrm{T}} Q_{b_{jk}}\left(\hat{R}(\theta_i)\,b^i_{jk} - b_{jk}\right)\right). \tag{8.18}$$

Equation (8.18) holds if and only if $\hat{R}(\theta_i)\,b^i_{jk} - b_{jk} = 0$. Therefore, $\hat{b}_{jk} = b_{jk}$ is the unique equilibrium point of (8.3) and $\hat{R}(\theta_i) = R(\theta_i)$.

Select a Lyapunov function

$$V = \frac{1}{2}\left\|\hat{b}_{jk} - b_{jk}\right\|^2. \tag{8.19}$$

Take the derivative of $V$,

$$\begin{aligned}\dot{V} &= -\gamma\delta\left(\alpha + \beta\tfrac{\dot{\mu}}{\mu}\right)\left(\left(\hat{R}(\theta_i)\,b^i_{jk} - b_{jk}\right)^{\mathrm{T}} Q_{b_{jk}}\left(\hat{R}(\theta_i)\,b^i_{jk} - b_{jk}\right)\right)\\ &\le -2\gamma\delta\left(\alpha + \beta\tfrac{\dot{\mu}}{\mu}\right)\lambda_{Q_{b_{jk}}} V,\end{aligned} \tag{8.20}$$

where $\lambda_{Q_{b_{jk}}} \ge 0$ is the minimum eigenvalues of $Q_{b_{jk}}$.

From (8.20), one has $\dot{V} \le -\alpha_3 V - \beta_3\frac{\dot{\mu}}{\mu}V$, where $\alpha_3 = 2\gamma\delta\alpha\lambda_{Q_{b_{jk}}}$ and $\beta_3 = 2\gamma\delta\beta\lambda_{Q_{b_{jk}}}$.

According to Lemma 1.1, one can obtain that $V(t)\begin{cases}\le \mu^{-\beta_3}e^{-\alpha_3 t}V(0),\ t\in[0,T)\\ \equiv 0,\ t\in[T,\infty)\end{cases}$. Therefore, $\hat{b}_{jk}$ can converge to $b_{jk}$ at the specified time, i.e., $\hat{R}(\theta_i) \to R(\theta_i)$. The upper limit of specified convergence time is $2(i-3)T + T$ in theory.

Next, we will prove $\hat{c}_i$ converges to $c_i$.

The equilibrium points of (8.4) satisfies

$$0 = -\left(\alpha + \beta\frac{\dot{\mu}}{\mu}\right)\left(Q_{\hat{b}_{ij}}\left(\hat{c}_i - \hat{c}_j\right) + Q_{\hat{b}_{ik}}\left(\hat{c}_i - \hat{c}_k\right)\right). \tag{8.21}$$

Multiply $\left(\hat{c}_i - c_i\right)^{\mathrm{T}}$ on both sides of (8.21), one has

$$\begin{aligned} 0 &= -\left(\alpha + \beta\tfrac{\dot{\mu}}{\mu}\right)\left(\hat{c}_i - c_i\right)^{\mathrm{T}}\left(Q_{\hat{b}_{ij}}\left(\hat{c}_i - \hat{c}_j\right) + Q_{\hat{b}_{ik}}\left(\hat{c}_i - \hat{c}_k\right)\right) \\ &= -\left(\alpha + \beta\tfrac{\dot{\mu}}{\mu}\right)\left(\hat{c}_i - c_i\right)^{\mathrm{T}}\times \\ &\left(Q_{\hat{b}_{ij}}\left(\hat{c}_i - c_i + c_i - \hat{c}_j\right) + Q_{\hat{b}_{ik}}\left(\hat{c}_i - c_i + c_i - \hat{c}_k\right)\right). \end{aligned} \tag{8.22}$$

Based on the above analysis, we have $\hat{c}_j \to c_j, \hat{c}_k \to c_k$ when $t \to 2(i-3)T$. According to the convergence analysis of $\hat{R}(\theta_i)$, we know that $\hat{b}_{ij} \to b_{ij}, \hat{b}_{ik} \to b_{ik}$ when $t \to 2(i-3)T + T$. Then, (8.22) is equivalent to

$$0 = -\left(\alpha + \beta\frac{\dot{\mu}}{\mu}\right)\left(\hat{c}_i - c_i\right)^{\mathrm{T}}\left(Q_{b_{ij}} + Q_{b_{ik}}\right)\left(\hat{c}_i - c_i\right). \tag{8.23}$$

Equation (8.23) holds if and only if $\hat{c}_i - c_i = 0$, so (8.23) has a unique equilibrium point $\hat{c}_i = c_i$.

Select a Lyapunov function

$$V = \frac{1}{2}\left\|\hat{c}_i - c_i\right\|^2. \tag{8.24}$$

Take the derivative of $V$

$$\begin{aligned} \dot{V} &= -\left(\alpha + \beta\tfrac{\dot{\mu}}{\mu}\right)\left(\hat{c}_i - c_i\right)^{\mathrm{T}}\left(Q_{b_{ij}} + Q_{b_{ik}}\right)\left(\hat{c}_i - c_i\right) \\ &\leq -2\left(\alpha + \beta\tfrac{\dot{\mu}}{\mu}\right)\lambda_{Q_{b_{ij}}+Q_{b_{ik}}} V, \end{aligned} \tag{8.25}$$

where $\lambda_{Q_{b_{ij}}+Q_{b_{ik}}}$ is the minimum eigenvalue of $Q_{b_{ij}} + Q_{b_{ik}}$ and $\lambda_{Q_{b_{ij}}+Q_{b_{ik}}} \geq 0$.

Inequality (8.25) can be reformulated as

$$\dot{V} \leq -\alpha_4 V - \beta_4\frac{\dot{\mu}}{\mu}V, \tag{8.26}$$

where $\alpha_4 = 2\alpha\lambda_{Q_{b_{ij}}+Q_{b_{ik}}}$ and $\beta_4 = 2\beta\lambda_{Q_{b_{ij}}+Q_{b_{ik}}}$. According to Lemma 1.1, one can obtain that $V(t)\begin{cases} \leq \mu^{-\beta_4}e^{-\alpha_4 t}V(0), t \in [0, T) \\ \equiv 0, t \in [T, \infty) \end{cases}$. Then, one can conclude that $\hat{c}_i$ can converge to $c_i$ at the specified time.

Combined with the convergence time $2(i-3)T + T$ of $\hat{R}(\theta_i)$, one can obtain that the estimated orientation information and estimated position information of follower agent $i$ converge to the real value at the specified time. The upper limit of specified convergence time is $2(i-2)T$ in theory.

The proof of Part II is completed.

Based on the analysis of Part I and Part II, one can draw the conclusion that after applying the robust prescribed-time distributed estimation algorithms (8.3) and (8.4), the estimated rotation matrix $\hat{R}(\theta_i)$ can converge to the real rotation matrix $R(\theta_i)$ and the estimated position $\hat{c}_i$ can converge to the global absolute position $c_i$ at the specified time.

The proof of Theorem 8.1 is completed.

## 8.4 Simulation and Experiment

In this section, the numerical example and the UAV experiment are given to illustrate the effectiveness and the realizability of the proposed methods.

### *8.4.1 Numerical Simulation*

In this subsection, some simulation results under different cases are provided. The global absolute positions of all leaders can be obtained directly. The position information of all followers should be estimated based on the measured local bearing information and the communication with their neighbors.

**Case 1** The bounded measurement noises

At first, a MASs with two leader agents and eight follower agents is considered. The orientations of all follower agents are given randomly, and the communication topology satisfies the Henneberg construction. As shown in Fig. 8.2, the red circles denote the two leaders. The blue circles are the follower agents, and the black vertical lines indicate the local reference frame, which is not align with the global one.

The measurement noise following Gauss distribution, the mathematical expectation is zero and the variance is $\frac{b_{jk}^i}{10}$. Select the parameter $h = 5$, $\lambda = 0.1$, $\alpha = 0.2$, $\beta = 5$ and set the upper limit of the specified convergence time $T = 4$. The initial estimated values are given randomly. Applying (8.3) and (8.4), the results are shown in Figs. 8.2, 8.3 and 8.4.

The convergence trajectories are shown in Fig. 8.2, where the green triangles are the initial estimated values and the green curves are the convergence trajectories. Figure 8.3 gives the orientation estimation error curves of all follower agents. One can find that the estimation error $\left\| \hat{R}(\theta_i) - R(\theta_i) \right\|$ can converge to near zero at time $T = 4$. The position estimation error curves are shown in Fig. 8.4, it can be seen that the estimation error $\left\| \hat{c}_i - c_i \right\|$ can converge to near zero at the fourth second.

Note that, the orientation estimation and the position estimation are performed simultaneously. This coupling causes most agents position estimates to diverge first and then converge as the orientation estimates converge gradually.

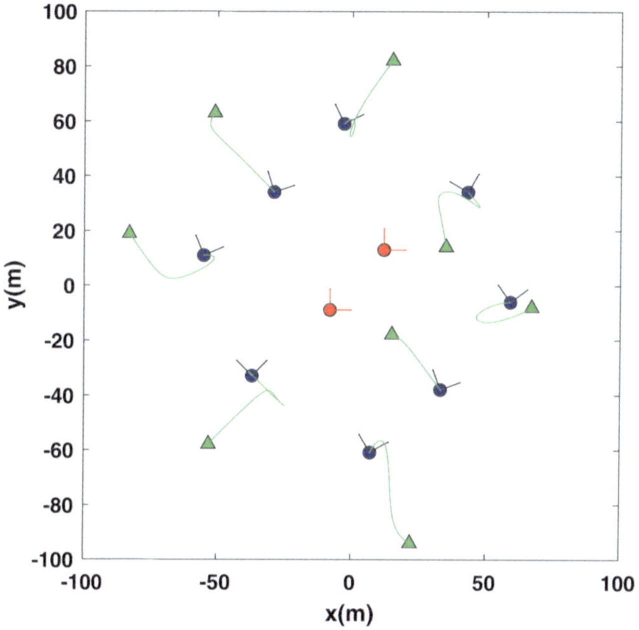


**Fig. 8.2** The convergence trajectories

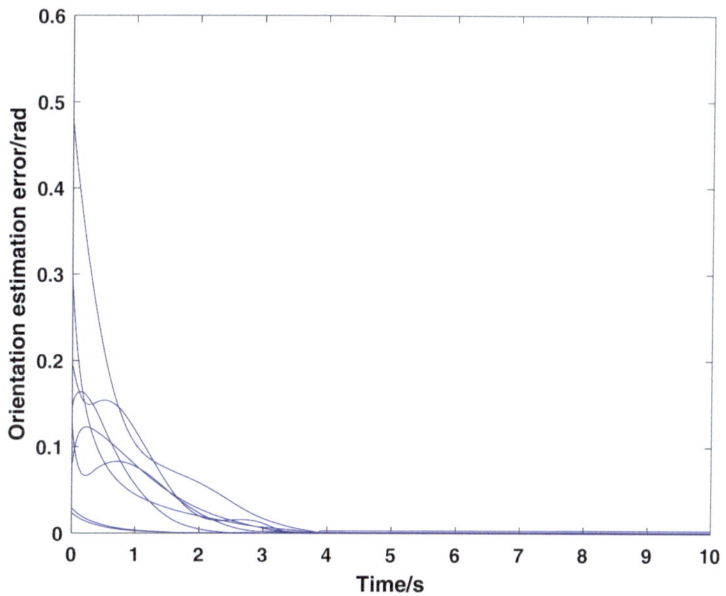


**Fig. 8.3** The orientation estimation error curves under measurement noise

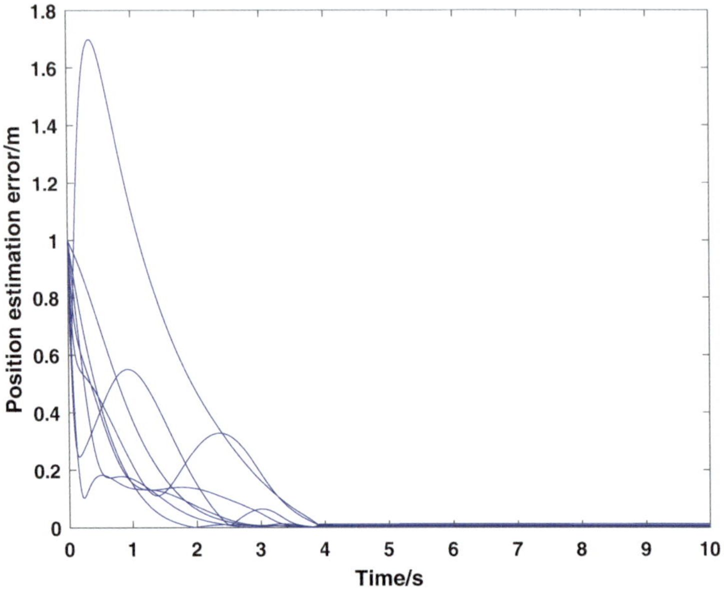


**Fig. 8.4** The position estimation error curves under measurement noise

Furthermore, in order to demonstrate the advantages of the proposed estimation algorithms (8.3) and (8.4). The continues-time distributed localization estimation scheme using the local bearing measurement is introduced.

$$\dot{\hat{b}}_{jk} = -\gamma Q_{\tilde{b}_{jk}} \hat{R}(\theta_i) \hat{b}^i_{jk}. \tag{8.27}$$

$$\dot{\hat{c}}_i = -Q_{\hat{b}_{ij}} \left(\hat{c}_i - \hat{c}_j\right) - Q_{\hat{b}_{ik}} \left(\hat{c}_i - \hat{c}_k\right). \tag{8.28}$$

The measurement noise and the parameters remain unchanged, applying (8.27) and (8.28), the simulation results are shown in Figs. 8.5 and 8.6. From Fig. 8.5, it can be seen that the orientation estimation error can gradually converge to zero, the convergence time is approximately 100 s. From Fig. 8.6, one can find that the position estimation error can converge to zero at the 140th s.

Compared Figs. 8.3 and 8.4 with Figs. 8.5 and 8.6, one can find that although the estimation error can gradually converge to zero after applying distributed estimation algorithms (8.27) and (8.28), the convergence speed is slower than after applying distributed estimation algorithms (8.3) and (8.4). Both the orientation estimation error and the position estimation error after applying algorithms (8.3) and (8.4) can quickly converge to an acceptable range.

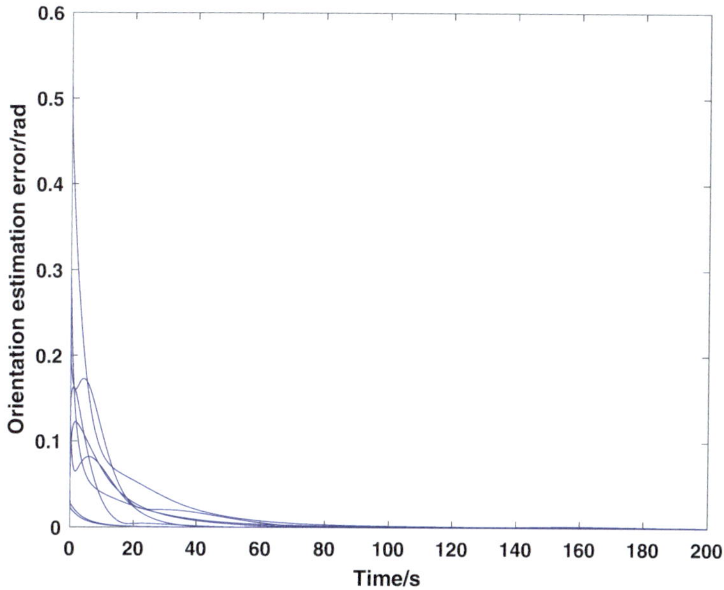


**Fig. 8.5** The orientation estimation error curves under measurement noise

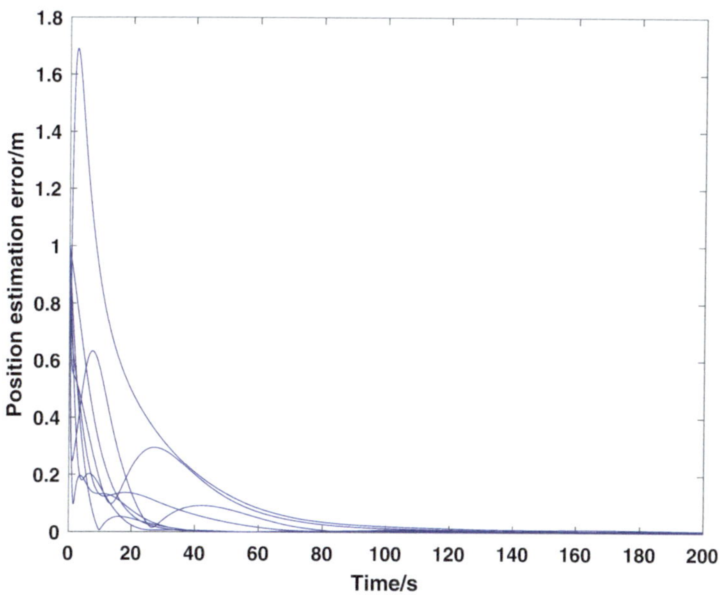


**Fig. 8.6** The position estimation error curves under measurement noise

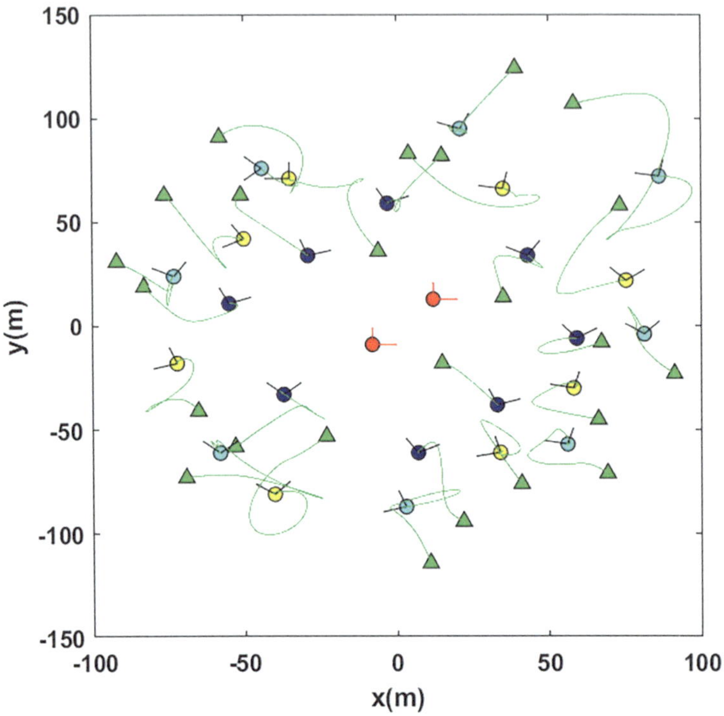


**Fig. 8.7** The convergence trajectories

Further, in order to test the scalability of the estimation algorithms (8.3) and (8.4). According to Henneberg configuration, the number of agents is increased to 26, where the cyan and yellow circles are the newly added agents. The parameters remain unchanged, the simulation results are shown in Figs. 8.7, 8.8 and 8.9, it can be seen that the MASs can still converges within the prescribed-time interval, and both the orientation estimation error and the position estimation error converge to near zero. Therefore, one can conclude that the proposed robust prescribed-time distributed localization method can be applied to large-scale MASs.

**Case 2** The ideal environment

In this case, the effectiveness of (8.3) and (8.4) is verified in the absence of measurement noise. Set constant parameter $\lambda = 0$ and the other parameters remain unchanged, the simulation results are shown in Figs. 8.10 and 8.11.

From Figs. 8.10 and 8.11, one can find that both the orientation estimation errors and the position estimation errors can converge to zero. Then, one can conclude that the proposed prescribed-time distributed estimation algorithms (8.3) and (8.4) can ensure the accurate orientation and position estimation under ideal environment.

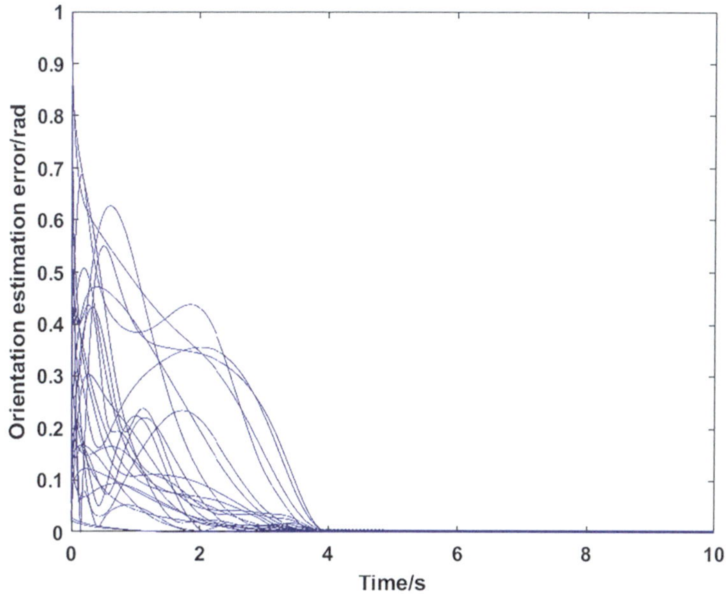


**Fig. 8.8** The orientation estimation error curves under measurement noise

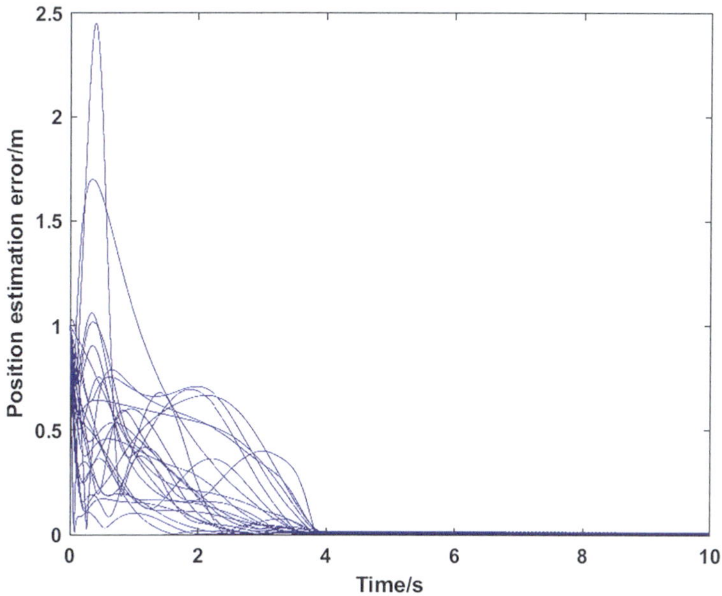


**Fig. 8.9** The position estimation error curves under measurement noise

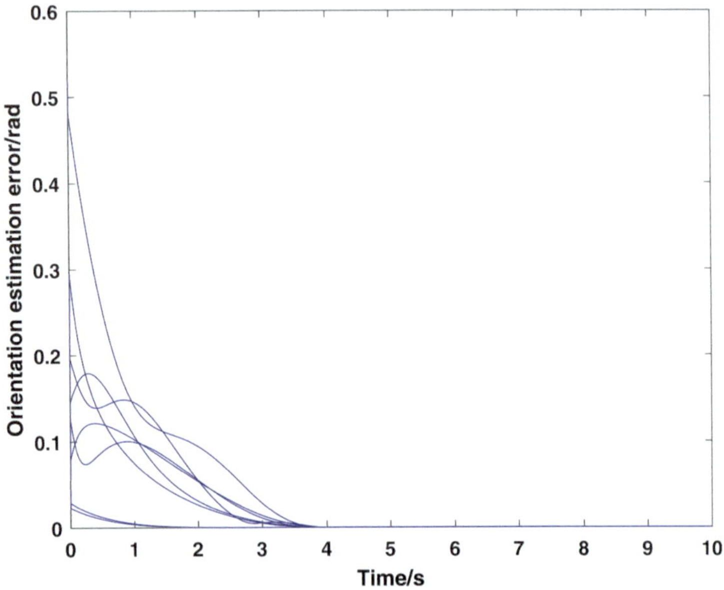


**Fig. 8.10** The orientation estimation error curves under ideal environment

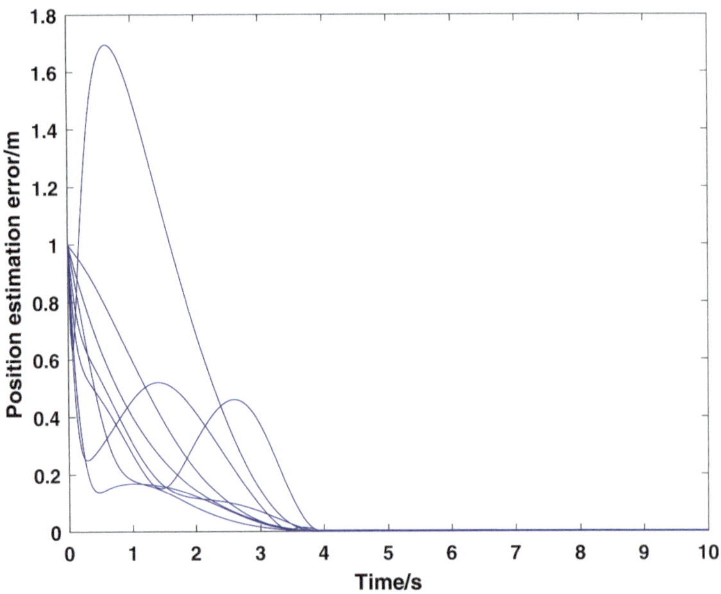


**Fig. 8.11** The position estimation error curves under ideal environment

### 8.4.2 Quad-Rotor UAV Experiment

In this subsection, the mini quad-rotor UAV experiment is provided to verify the practicability of (8.3) and (8.4).

OptiTrack system is a high-precision optical motion capture system. In this experiment, ten mini quad-rotor UAV crazyflies with passive spherical markers are used to test the validity of (8.3) and (8.4).

Figure 8.12 shows the deployment of the ten UAVs. The position information of UAV is obtained by the OptiTrack system, the leaders know their global absolute positions, and the followers need to estimate their position information. In the experiment, the OptiTrack system is used to measure the local bearing information.

The initial estimated values of the follower UAVs are given randomly. Due to the presence of measurement noise in the OptiTrack system when measuring local bearing information, no additional measurement noise is added. Select $h = 3$, $\lambda = 0.15$, $\alpha = 4$, $\beta = 5$ and set the upper limit of the specified convergence time $T = 4$. Applying algorithms (8.3) and (8.4), the experiment results are shown in Figs. 8.13, 8.14 and 8.15.

The convergence trajectories are shown Fig. 8.13, where two leader UAVs are marked by red stars and the rest 8 follower UAVs are marked by blue circles. Each follower UAV has different orientation, and the direction of the local reference frame is represented by red-green-blue lines. Note that each UAVs z-axis is aligned with the z-axis of the global reference frame. The black arrow indicates that the UAV can measure the local bearing information of its neighbors and receive position estimation information from its neighbors. Figure 8.14 shows the orientation estimation

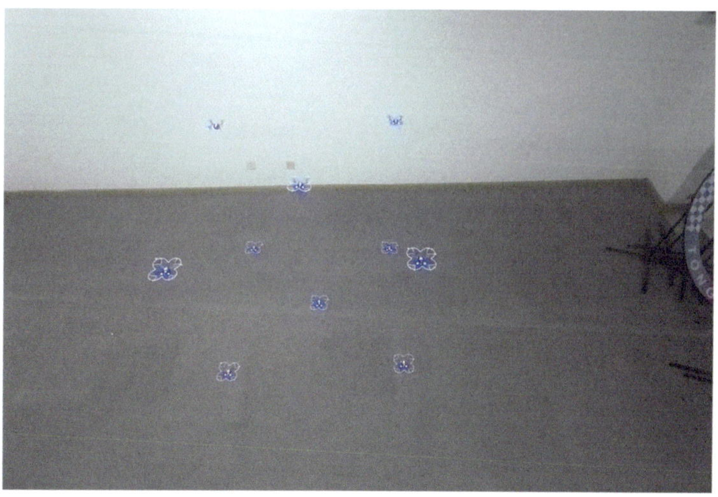

**Fig. 8.12** The deployment of the ten UAVs

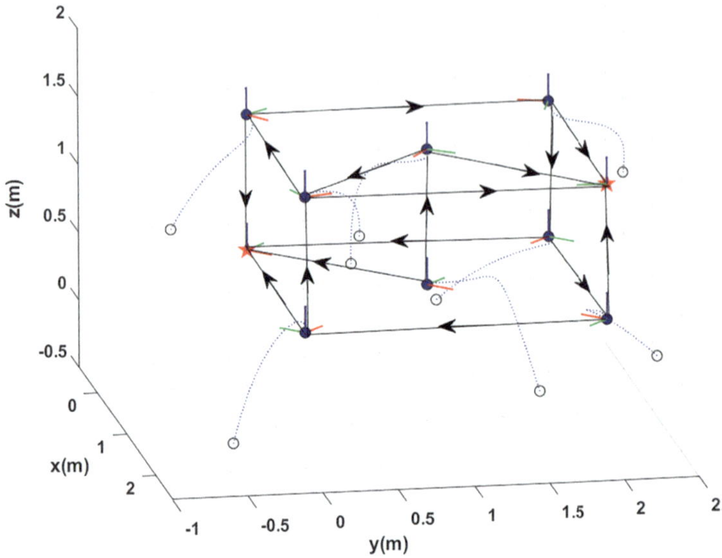


**Fig. 8.13** The convergence trajectories

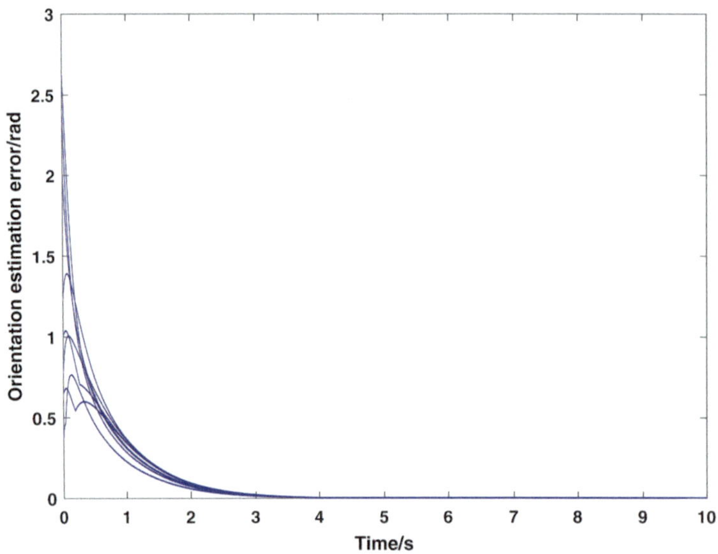


**Fig. 8.14** The orientation estimation error curves

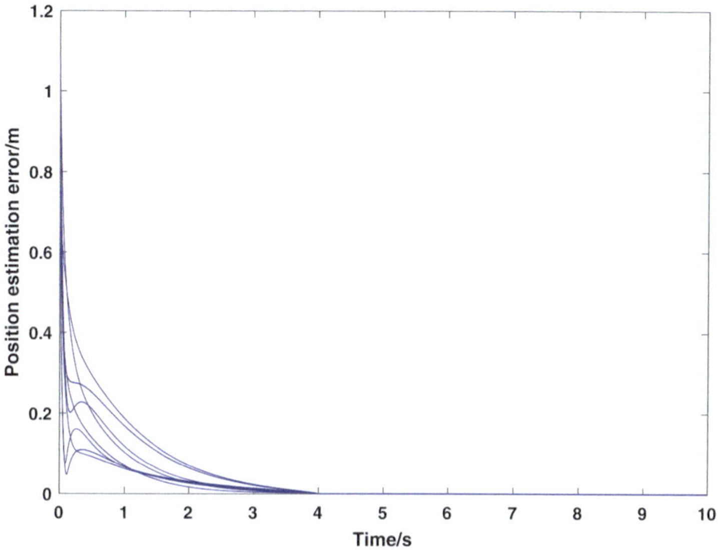


**Fig. 8.15** The position estimation error curves

error curves of all follower UAVs. One can find that the estimation error can converge to near zero at time $T = 4$. The position estimation error curves are shown in Fig. 8.15, it can be seen that the position estimation error can converge to near zero at the fourth second.

The effectiveness and the realizability of the proposed distributed localization estimation schemes are verified through the simulation and experiment results. Then, one can draw the conclusions that: (a) The proposed robust prescribed-time distributed localization estimation scheme can realize the orientation alignment and the global absolute position localization under bounded measurement noise. (b) The proposed estimation scheme can also applicable to large-scale MASs.

## 8.5 Conclusions

In this chapter, a local-bearing-based robust prescribed-time distributed localization estimation scheme is proposed for MASs with bounded measurement noise under directed graph. Each follower agent can align its orientation angle with the global reference frame and estimate its global absolute position. The designed unbiased estimator and the time-varying gain can suppress the measurement noise. Based on cascade system, the global convergence of the proposed estimator is proved by introducing mathematical induction. The simulation and experiment results also prove the effectiveness of the proposed estimation algorithms. The proposed localization scheme can be applied to environmental monitoring, field search and rescue, and other scenarios.

## References

1. Oh, K., Ahn, H.: Formation control and network localization via orientation alignment. IEEE Trans. Autom. Control **59**(2), 540–545 (2014)
2. Trinh, M., Lee, B., Ye, M., et al.: Bearing-based formation control and network localization via global orientation estimation. In: Proceedings of IEEE Conference on Control Technology and Applications, pp. 1084–1089. Copenhagen, , Denmark (2018)
3. Barooah, P., Hespanha, J.: Estimation on graphs from relative measurements. IEEE Control Syst. Mag. **27**(4), 57–74 (2007)
4. Lin, Z., Fu, M., Diao, Y.: Distributed self-localization for relative position sensing networks in 2D space. IEEE Trans. Signal Process. **63**(14), 3751–3761 (2015)
5. Polyakov, A.: Nonlinear feedback design for fixed-time stabilization of linear control systems. IEEE Trans. Autom. Control **57**(8), 2106–2110 (2012)
6. Zuo, Z., Tie, L.: A new class of finite-time nonlinear consensus protocols for multi-agent systems. Int. J. Control **87**(2), 363–370 (2014)
7. Song, Y., Wang, Y., Holloway, J., et al.: Time varying feed-back for regulation of normal form nonlinear systems in prescribed finite time. Automatica **83**, 243–251 (2017)
8. Li, X., Zhu, Y., Zhao, X., et al.: Bearing based prescribed time formation tracking for second order multi agent systems. IEEE Trans. Circ. Syst. II: Express Briefs **69**(7), 3259–3263 (2022)

# Chapter 9
# Integration of Distributed Localization and Formation of MASs

This chapter investigates the localizability and controllability of local-bearing-based multi-agent systems under directed graph and develops the integrated distributed localization and formation scheme. Based on local-bearing measurement, a novel prescribed-time orientation estimation algorithm is designed to ensure that the local reference frame of the follower agent is aligned with the global one. The local bearing information is more easily obtained than global one. Therefore, the new orientation estimation result is expected to be more widely applicable. The orientation estimation result is then applied to the stationary and moving target formation with unknown initial positions, two prescribed-time distributed formation control schemes are developed. The distinctive advantage of this work is that only the local bearing information is used, the fast and controllable localization estimation and formation control are achieved simultaneously. The global convergence is derived based on the cascade system and the mathematical induction method. Both simulation and experiment results are provided to prove the effectiveness of the theoretical results.

## 9.1 Introduction

Formation control, as an important research area in multi-agent cooperative control, has been extensively studied over the past few decades. Existing multi-agent formation control methods can be divided into two categories based on the desired geometric configuration of the target formation: formations defined by the relative positions between neighbors or by relative distances [1, 2]. Methods based on relative positions can achieve translational maneuvers of the formation, while methods based on relative distances can achieve both translational and rotational maneuvers. In recent years, with the development of visual sensors and wireless sensor arrays, formation control methods based on bearing measurements have garnered significant attention from experts and scholars [3–5].

 
Y. Lv et al., *Distributed Localization and Formation Control of Multi-Agent Systems with Local Sensing Information*, https://doi.org/10.1007/978-981-96-6015-5_9

In bearing-based formation control algorithms, the geometric configuration of the target formation is defined by the desired relative bearing angles between adjacent agents. Since relative bearing angles are relatively easy to obtain, control algorithms based solely on bearing angles offer potential solutions for achieving formation control using onboard sensors in multi-agent systems. Based on parallel rigidity theory, Bishop et al. [6] introduced the bearing rigidity matrix and proposed a control algorithm with local asymptotic stability, but this algorithm still required the measurement of relative positions. By introducing a decentralized position estimator, Giordano et al. [7] eliminated this requirement. Zhao and Zelazo [8] extended bearing rigidity theory to arbitrary dimensions and proposed an infinitesimal bearing rigidity formation control algorithm with almost global asymptotic stability. Furthermore, Zhao et al. [9] introduced the bearing Laplacian matrix to characterize the algebraic properties of bearing rigidity structures. Building on the aforementioned works, Ding et al. [10] designed a novel purely bearing-based formation control algorithm that ensures global exponential convergence.

Despite significant progress in purely bearing-based formation control, many important issues remain unresolved due to the highly nonlinear dynamics of formations. Existing bearing-based formation control algorithms are proposed under the assumption that a global reference frame is known, and the control algorithms use global bearing angle information. However, in practical applications, each agent has its own local reference frame, and obtaining local bearing angle information is much easier and more practical than acquiring global bearing angle information. Therefore, distributed localization estimation and formation control for multi-agent systems under local reference frames remain open topics worthy of in-depth research.

Additionally, multi-agent system formation control often has certain requirements regarding task time, making convergence time an important performance indicator. Finite-time control has been systematically studied in multi-agent systems to achieve faster convergence speeds. Zhao et al. [11] used sign functions to suppress relative bearing errors, achieving finite-time convergence. References [12, 13] achieved almost global convergence of infinitesimal bearing rigidity formations in finite time. In recent studies, Zhao and Ding et al. extended these results to specified-time convergence [14]. All the aforementioned work was conducted under the condition that the global reference frame is known. The problem of achieving fast convergence in formation control for multi-agent systems under local reference frames has not yet been studied.

Motivated by the above discussion and analysis, this chapter studies the localizability and controllability of MASs under directed graph, develops a novel local-bearing-based prescribed-time integrated localization estimation and formation control scheme in a distributed manner. Different from the existing works, each agent can only measure the bearing information of its neighbors within its local reference framework, which is called local bearing information. Except for the leaders, the position information of all followers is unknown.

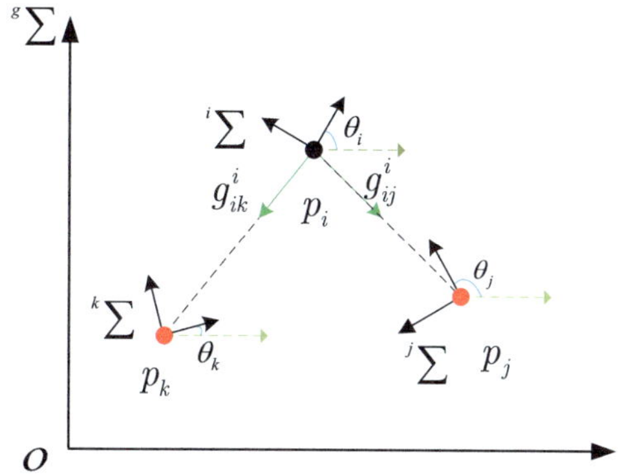


**Fig. 9.1** The local bearing measurement

The main contributions of this work are summarized as follows:

(i) A novel prescribed-time orientation estimation method is designed, which ensures the local reference frame is aligned with the global one.
(ii) A prescribed-time distributed position estimation method merely using local bearing information is proposed, which removes the constraint of initial orientation angles and the relative orientation measurements in [15–17].
(iii) The prescribed-time distributed formation control scheme based on local bearing measurement combined with orientation and position estimation is developed, which realizes the stationary and moving target formation for MASs in local reference frame.

The rest of the paper is organized as follows: Sect. 9.2 presents the problem formulation. Sections 9.3 and 9.4 shows the local-bearing-only stationary and moving target formation control of single-integrator, respectively. Section 9.5 gives the simulation and experimental results. Section 9.6 concludes this work.

## 9.2 Problem Formulation

Consider a $N$-agents MASs in two-dimensional plane, where two agents are leaders (labelled as 1 and 2) and the remaining $(N-2)$ are followers (labelled as $\{3, \ldots, N\}$). In this paper, the follower agent has its own local reference frame, which is not aligned with the global reference frame $^g\Sigma$, refer to Fig.9.1 for details.

Let $^i\Sigma$ represents the local reference frame of the follower agent $i$ and $\theta_i \in [0, 2\pi]$ denotes the orientation angle with respect to $^g\Sigma$. Define $g_{ik} = \frac{p_k - p_i}{\|p_k - p_i\|}$ is the global unit direction vector, where $p_k$ and $p_i$ are the global absolute positions of $k$ and $i$. $g^i_{ik}$ is the relative bearing vector of agent $k$ with respect to agent $i$ under the local reference frame $^i\Sigma$.

**Assumption 9.1** The communication topology graph of the considered multi-agent systems satisfies the Henneberg construction.

The desired target formation that the agents are expected to accomplish is defined as follows.

**Definition 9.1** (Stationary target formation)
The stationary target formation $(\mathcal{G}, p^*)$ satisfies the time-invariant desired neighboring bearings $\left\{g_{ij}^*\right\}_{(i,j)\in\mathcal{E}}$ and the time-invariant global coordinates $p_j^*$ of the leader agent $j$.

**Definition 9.2** (Moving target formation)
The moving target formation $(\mathcal{G}, p^*(t))$ satisfies the time-invariant desired neighboring bearings $\left\{g_{ij}^*\right\}_{(i,j)\in\mathcal{E}}$ and the time-varying global coordinates $p_j^*(t)$ of the leader agent $j$. The leaders move at a common constant velocity.

**Remark 9.1** The stationary and the moving target formations in definitions 9.1 and 9.2 are jointly determined by the desired time-invariant bearings and the leader agents. In this chapter, we only consider the desired bearings and the leader positions that ensure the presence of the target formation. Feasible bearings and leader positions can be obtained from a formation with the desired geometric pattern. Meanwhile, the desired bearings and leader positions fulfill certain conditions to ensure uniqueness of the target formation, refer [10] for details. Note that, the size of the formation can be controlled by repositioning the leader.

In this chapter, each follower agent can measure the bearing information of its neighbors, i.e., the local bearing information, and receive the position information transmitted by the neighbors. The considered agent is modeled by a single-integrator dynamic model,

$$(p_i)' = u_i, i = 1, \ldots, N$$

where $p_i = [x_i, y_i] \in \mathbb{R}^2$ represents the global coordinates of agent $i$, and $u_i$ is the control input to be designed.

The objective of this chapter is to design a novel local-bearing-based prescribed-time integrated localization estimation and formation control scheme, which can ensure the orientation alignment and the initial position estimation, and realize the desired target formation, i.e., $\lim_{t\to T} \hat{R}(\theta_i) = R(\theta_i)$, $\lim_{t\to T} \hat{p}_i^0 = p_i^0$ and $\lim_{t\to T} p_i(t) = p_i^*(t)$, where $T$ is the prescribed convergence time, $\hat{R}(\theta_i)$ is the estimated rotation matrix, $\hat{p}_i^0$ denotes the estimated initial position information and $p_i^0$ is the actual initial position information, $p_i(t)$ is the real-time position information and $p_i^*(t)$ represents the desired position information.

## 9.3 Local-Bearing-Only Stationary Target Formation Tracking Control

This section studies the local-bearing-based formation tracking control of MASs with single-integrator dynamic. The initial position information $p_i^0$ and the orientation angle $\theta_i$ of the follower agent $i$ is unknown, only the local bearing information is available. Section 9.3.1 investigates the stationary target formation and proposes an integrated localization estimation and formation control scheme. Section 9.3.2 provides the convergence analysis.

### 9.3.1 *Stationary Target Formation*

For the follower agent $i$, the integrated prescribed-time distributed localization and formation control scheme is designed as follows

$$(\hat{g}_{jk})' = -\left(a + b\frac{\mu'}{\mu}\right) Q_{\tilde{g}_{jk}} \hat{R}(\theta_i)\, g_{jk}^i, \tag{9.1}$$

$$(\hat{p}_i^0)' = -\left(a + b\frac{\mu'}{\mu}\right) \left(Q_{\hat{g}_{ij}} (\hat{p}_i^0 - \hat{p}_j^0) + Q_{\hat{g}_{ik}} (\hat{p}_i^0 - \hat{p}_k^0)\right), \tag{9.2}$$

$$u_i = -\left(a + b\frac{\mu'}{\mu}\right) \left(Q_{g_{ij}^*} (-\hat{g}_{ij}) + Q_{g_{ik}^*} (-\hat{g}_{ik})\right), \tag{9.3}$$

where $\hat{g}_{jk}$ is the estimated global unit vector between agents $j$ and $k$, $\hat{p}_i^0$ is the estimated initial position information of agent $i$, $u_i$ is the designed control input. $g_{jk}^i = \frac{g_{ik}^i - g_{ij}^i}{\left\| g_{ik}^i - g_{ij}^i \right\|}$, $g_{ik}^i$ and $g_{ij}^i$ are the local bearing information that can be directly measured in the local reference frame of agent $i$. $\tilde{g}_{jk} = \frac{\hat{p}_j^0 - \hat{p}_k^0}{\left\| \hat{p}_j^0 - \hat{p}_k^0 \right\|}$, $\hat{p}_j^0$ and $\hat{p}_k^0$ are the estimated initial position information of the agents $j$ and $k$. $\hat{g}_{ij} = \hat{R}(\theta_i)\, g_{ij}^i$, $\hat{g}_{ik} = \hat{R}(\theta_i)\, g_{ik}^i$. $a, b \in \mathbb{R}_{>0}$, $g_{ij}^*$ and $g_{ik}^*$ are the time-invariant desired neighboring bearings.

The estimation algorithm (9.1) is used to align the orientation, i.e., estimate the orientation angle $\theta_i$. The estimation method (9.2) is designed to estimate the initial position information of the follower agent. Since the estimation of orientation angle $\theta_i$ depends on the estimation of the initial position, only the global initial position estimation of the followers is needed. The control algorithm (9.3) is proposed to realize the desired stationary target formation, the position of the followers can be uniquely determined from the leader's global position information and the given desired bearing information.

To better explain the orientation estimation algorithm (9.1), a lemma is provided to show that the orientation angle $\theta_i$ can be estimated indirectly using the estimated unit direction vector $\hat{g}_{jk}$.

**Lemma 9.1** *Based on local bearing information, the proposed distributed orientation estimation algorithm (9.1) can realize indirect estimation of the orientation angle $\theta_i$ through the estimation of global unit direction vector $\hat{g}_{jk}$.*

***Proof*** In this work, the local bearing information can be directly measured. Since $\hat{g}_{jk} = \hat{R}(\theta_i)\, g^i_{jk}$, one can obtain that $(\hat{g}_{jk})' = \left(\hat{R}(\theta_i)\right)' g^i_{jk}$.

Define $(\hat{g}_{jk})' = \left((\hat{\varphi}_x)', (\hat{\varphi}_y)'\right)^{\mathrm{T}}$, $\hat{g}_{jk} = (\hat{\varphi}_x, \hat{\varphi}_y)^{\mathrm{T}}$ and $g^i_{jk} = (\varphi^i_x, \varphi^i_y)^{\mathrm{T}}$, then one has

$$\hat{R}(\theta_i) = \begin{bmatrix} \varphi^i_x\hat{\varphi}_x + \varphi^i_y\hat{\varphi}_y & \hat{\varphi}_x\varphi^i_y - \varphi^i_x\hat{\varphi}_y \\ -\hat{\varphi}_x\varphi^i_y + \varphi^i_x\hat{\varphi}_y & \varphi^i_y\hat{\varphi}_y + \varphi^i_x\hat{\varphi}_x \end{bmatrix},$$

$$\left(\hat{R}(\theta_i)\right)' = \begin{bmatrix} \varphi^i_x(\hat{\varphi}_x)' + \varphi^i_y(\hat{\varphi}_y)' & (\hat{\varphi}_x)'\varphi^i_y - \varphi^i_x(\hat{\varphi}_y)' \\ -(\hat{\varphi}_x)'\varphi^i_y + \varphi^i_x(\hat{\varphi}_y)' & \varphi^i_y(\hat{\varphi}_y)' + \varphi^i_x(\hat{\varphi}_x)' \end{bmatrix}.$$

Based on the above analysis, one can conclude that the orientation angle $\theta_i$ can be estimated indirectly by the estimated unit direction vector $\hat{g}_{jk}$.

The proof of Lemma 9.1 is completed.

In order to better describe the proposed integrated prescribed-time distributed localization estimation and formation control scheme (9.1)–(9.3), a detailed block diagram is given in Fig. 9.2.

### 9.3.2 Convergence Analysis

The convergence of the integrated prescribed-time distributed localization estimation and formation control scheme (9.1)–(9.3) is shown below.

**Theorem 9.1** *Suppose the MASs satisfies Assumption 9.1, applying the integrated prescribed-time distributed localization estimation and formation control scheme (9.1)–(9.3), one can ensure that the estimated rotation matrix $\hat{R}(\theta_i)$ converges to the real rotation matrix $R(\theta_i)$, the estimated initial global position $\hat{p}^0_i$ converges to the global absolute position $p^0_i$ and the follower agents move to the desired position at the specified time.*

*In this section, the mathematical induction is introduced to prove Theorem 9.1. The follower agent $i \in \{3, \ldots, N\}$ has two neighbors $j$ and $k$, $1 \le j \ne k \le i-1$. Note that, the neighbors of follower agent 3 are leader agents 1 and 2.*

***Proof*** The proof is mainly divided into two parts, the specific is given below.

In the first part, we will prove that $\hat{R}(\theta_3)$ converges to $R(\theta_3)$, $\hat{p}^0_3$ converges to $p^0_3$, and $p_3$ move to $p^*_3$. In the second part, we will prove that $\hat{R}(\theta_i)$ converges to $R(\theta_i)$, $\hat{p}^0_i$ converges to $p^0_i$, and $p_i$ move to $p^*_i$.

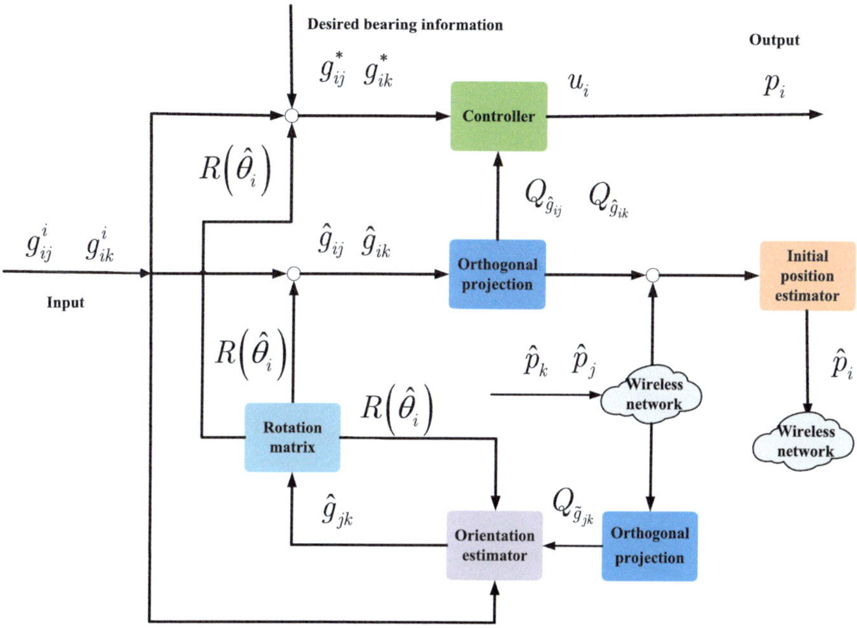


**Fig. 9.2** The block diagram of the integrated distributed localization estimation and formation control scheme

**Part I**. The convergence of the estimated rotation matrix $\hat{R}(\theta_3)$, the estimated initial position $\hat{p}_3^0$ and the actual global position $p_3$.

**Step 1**. The convergence of the estimated rotation matrix $\hat{R}(\theta_3)$

For follower agent 3, the equilibrium points of (9.1) satisfies

$$0 = -\left(a + b\frac{\mu'}{\mu}\right) Q_{g_{12}} R\left(\hat{\theta}_3\right) g_{12}^3. \tag{9.4}$$

Multiply $\left(\hat{g}_{12} - g_{12}\right)^{\mathrm{T}}$ on both sides of (9.4), one has

$$\begin{aligned} 0 &= -\left(a + b\tfrac{\mu'}{\mu}\right)\left(\hat{g}_{12} - g_{12}\right)^{\mathrm{T}} Q_{g_{12}} R\left(\hat{\theta}_3\right) g_{12}^3 \\ &= -\left(a + b\tfrac{\mu'}{\mu}\right)\left(R\left(\hat{\theta}_3\right) g_{12}^3 - g_{12}\right)^{\mathrm{T}} Q_{g_{12}} \left(R\left(\hat{\theta}_3\right) g_{12}^3 - g_{12} + g_{12}\right). \end{aligned} \tag{9.5}$$

Since $Q_{g_{12}} g_{12} = 0$, (9.5) becomes

$$0 = -\left(a + b\tfrac{\mu'}{\mu}\right)\left(R\left(\hat{\theta}_3\right) g_{12}^3 - g_{12}\right)^{\mathrm{T}} Q_{g_{12}} \left(R\left(\hat{\theta}_3\right) g_{12}^3 - g_{12}\right). \tag{9.6}$$

Equation (9.6) holds when $R\left(\hat{\theta}_3\right) g_{12}^3 - g_{12} = 0$. In other words, the orientation alignment algorithm (9.1) has a unique equilibrium point $\hat{g}_{12} = g_{12}$.

Substitute $g_{12}^3 = R\left(\hat{\theta}_3\right)^{-1} g_{12}$ into $R\left(\hat{\theta}_3\right) g_{12}^3 - g_{12} = 0$, one can obtain that

$$\left(R\left(\hat{\theta}_3\right) - R\left(\theta_3\right)\right) g_{12} = 0. \tag{9.7}$$

Denote $g_{12}=[\varphi_1, \varphi_2]^{\mathrm{T}}$, $R\left(\hat{\theta}_3\right) = \begin{bmatrix} \cos\hat{\theta}_3 & -\sin\hat{\theta}_3 \\ \sin\hat{\theta}_3 & \cos\hat{\theta}_3 \end{bmatrix}$, $R\left(\theta_3\right) = \begin{bmatrix} \cos\theta_3 & -\sin\theta_3 \\ \sin\theta_3 & \cos\theta_3 \end{bmatrix}$, then (9.7) becomes

$$\begin{bmatrix} -2\sin\frac{\hat{\theta}_3+\theta_3}{2}\sin\frac{\hat{\theta}_3-\theta_3}{2} & -2\cos\frac{\hat{\theta}_3+\theta_3}{2}\sin\frac{\hat{\theta}_3-\theta_3}{2} \\ 2\cos\frac{\hat{\theta}_3+\theta_3}{2}\sin\frac{\hat{\theta}_3-\theta_3}{2} & -2\sin\frac{\hat{\theta}_3+\theta_3}{2}\sin\frac{\hat{\theta}_3-\theta_3}{2} \end{bmatrix} [\varphi_1, \varphi_2]^{\mathrm{T}} = 0. \tag{9.8}$$

Since $g_{12}$ is a unit vector, $\varphi_1$ and $\varphi_2$ are not equal to zero at the same time (such as $\varphi_1 = 0$, $\varphi_2 = 1$ or $\varphi_1 = 1$, $\varphi_2 = 0$). From (9.8), one has $\sin\frac{\hat{\theta}_3+\theta_3}{2}\sin\frac{\hat{\theta}_3-\theta_3}{2} = 0$ and $\cos\frac{\hat{\theta}_3+\theta_3}{2}\sin\frac{\hat{\theta}_3-\theta_3}{2} = 0$. Due to the fact that $\hat{\theta}_3, \theta_3 \in [0, 2\pi]$, the necessary and sufficient condition for (9.8) to hold is $\hat{\theta}_3 = \theta_3$, that is $R\left(\hat{\theta}_3\right) = R\left(\theta_3\right)$.

Based on the above analysis (9.5)–(9.8), one can conclude that the orientation alignment algorithm (9.1) has a unique equilibrium point $\hat{g}_{12} = g_{12}$, and $R\left(\hat{\theta}_3\right) = R\left(\theta_3\right)$.

Design a Lyapunov function

$$V_1 = \frac{1}{2}\left\|\hat{g}_{12} - g_{12}\right\|^2. \tag{9.9}$$

Based on the above analysis of (9.5) and (9.6), the derivative of $V_1$ becomes

$$\begin{aligned} (V_1)' &= -\left(a + b\frac{\mu'}{\mu}\right)\left(R\left(\hat{\theta}_3\right) g_{12}^3 - g_{12}\right)^{\mathrm{T}} Q_{g_{12}} \left(R\left(\hat{\theta}_3\right) g_{12}^3 - g_{12}\right) \\ &\le -\left(a + b\frac{\mu'}{\mu}\right)\lambda_{Q_{g_{12}}}\left\|\hat{g}_{12} - g_{12}\right\|^2 \\ &= -2\left(a + b\frac{\mu'}{\mu}\right)\lambda_{Q_{g_{12}}} V_1 \\ &= -2a\lambda_{Q_{g_{12}}} V_1 - 2b\lambda_{Q_{g_{12}}}\frac{\mu'}{\mu} V_1, \end{aligned} \tag{9.10}$$

where $\lambda_{Q_{g_{12}}}$ is the minimum eigenvalue of matrix $Q_{g_{12}}$. Since orthogonal projection matrix $Q_{g_{12}}$ is positive semidefinite, $\lambda_{Q_{g_{12}}} \ge 0$.

From (9.10), one has $(V_1)' \le -\alpha_1 V_1 - \beta_1\frac{\mu'}{\mu} V_1$, where $\alpha_1 = 2a\lambda_{Q_{g_{12}}}$, $\beta_1 = 2b\lambda_{Q_{g_{12}}}$. It then follows from Lemma 1.1 that $V_1(t) \begin{cases} \le \mu^{-\beta_1} e^{-\alpha_1 t} V_1(0), & t \in [0, T) \\ \equiv 0, & t \in [T, \infty) \end{cases}$, which implies that $\hat{g}_{12}$ can converge to $g_{12}$ at the prespecified time $T$, that is $R\left(\hat{\theta}_3\right)$ converges to $R\left(\theta_3\right)$.

**Step 2**. The convergence of the estimated initial position $\hat{p}_3^0$.

Note that the neighbors of follower agent 3 are leader agents 1 and 2. Then, the equilibrium points of (9.2) satisfies

$$0 = -\left(a + b\frac{\mu'}{\mu}\right)\left(Q_{\hat{g}_{31}}\left(\hat{p}_3^0 - p_1^0\right) + Q_{\hat{g}_{32}}\left(\hat{p}_3^0 - p_2^0\right)\right), \tag{9.11}$$

where $p_1^0$ and $p_2^0$ are the initial position of leaders 1 and 2, respectively. Note that the position of leader agent is static invariant.

Multiply $\left(\hat{p}_3^0 - p_3^0\right)^{\mathrm{T}}$ on both sides of (9.11), one has

$$\begin{aligned} 0 &= -\left(a + b\tfrac{\mu'}{\mu}\right)\left(\hat{p}_3^0 - p_3^0\right)^{\mathrm{T}}\left(Q_{\hat{g}_{31}}\left(\hat{p}_3^0 - p_1^0\right) + Q_{\hat{g}_{32}}\left(\hat{p}_3^0 - p_2^0\right)\right) \\ &= -\left(a + b\tfrac{\mu'}{\mu}\right)\left(\hat{p}_3^0 - p_3^0\right)^{\mathrm{T}} \times \\ &\left(Q_{\hat{g}_{31}}\left(\hat{p}_3^0 - p_3^0 + p_3^0 - p_1^0\right) + Q_{\hat{g}_{32}}\left(\hat{p}_3^0 - p_3^0 + p_3^0 - p_2^0\right)\right). \end{aligned} \tag{9.12}$$

Based on the convergence analysis of $R\left(\hat{\theta}_3\right)$, one has $\hat{g}_{31}$ can converge to $g_{31}$ and $\hat{g}_{32}$ can converge to $g_{32}$ when $t \to T$. Since $Q_{g_{ij}} g_{ij} = 0$, one can obtain that $Q_{g_{31}}\left(p_3^0 - p_1^0\right) = 0$ and $Q_{g_{32}}\left(p_3^0 - p_2^0\right) = 0$.

It then follows (9.12) that

$$0 = -\left(a + b\tfrac{\mu'}{\mu}\right)\left(\hat{p}_3^0 - p_3^0\right)^{\mathrm{T}}\left(Q_{g_{31}}\left(\hat{p}_3^0 - p_3^0\right) + Q_{\hat{g}_{32}}\left(\hat{p}_3^0 - p_3^0\right)\right). \tag{9.13}$$

Equation (9.13) holds if and only if $\hat{p}_3^0 = p_3^0$,

Select a Lyapunov function

$$V_2 = \frac{1}{2}\left\| \hat{p}_3^0 - p_3^0 \right\|^2. \tag{9.14}$$

Take the derivation of $V_2$ and based on the analysis from (9.11) to (9.13), one can obtain that

$$\begin{aligned} (V_2)' &= -\left(a + b\tfrac{\mu'}{\mu}\right)\left(\hat{p}_3^0 - p_3^0\right)^{\mathrm{T}}\left(Q_{g_{31}} + Q_{g_{32}}\right)\left(\hat{p}_3^0 - p_3^0\right) \\ &\le -2\left(a + b\tfrac{\mu'}{\mu}\right)\lambda_{Q_{g_{31}}+Q_{g_{32}}} V_2, \end{aligned} \tag{9.15}$$

where $\lambda_{Q_{g_{31}}+Q_{g_{32}}}$ represent the minimum eigenvalue of matrix $\left(Q_{g_{31}} + Q_{g_{32}}\right)$, and $\lambda_{Q_{g_{31}}+Q_{g_{32}}} \ge 0$.

From (9.15), one can obtain that $(V_2)' \le -\alpha_2 V_2 - \beta_2 \frac{\mu'}{\mu} V_2$, where $\alpha_2 = 2a\lambda_{Q_{g_{31}}+Q_{g_{32}}}$, $\beta_2 = 2b\lambda_{Q_{g_{31}}+Q_{g_{32}}}$.

According to Lemma 1.1, one has $V_2(t)\begin{cases} \le \mu^{-\beta_2}e^{-\alpha_2 t}V_2(0), & t \in [0, T) \\ \equiv 0, & t \in [T, \infty) \end{cases}$. Therefore, one can conclude that $\hat{p}_3^0$ can converge to $p_3^0$ at the prespecified time $2T$.

Based on the above analysis, one can draw the conclusion that the estimated rotation matrix $\hat{R}(\theta_3)$ and the estimated initial position $\hat{p}_3^0$ can converge to the actual value at the prespecified time. The upper limit of specified convergence time is $2T$ in theory.

**Step 3**. The convergence of $p_3$.

Next, we will prove that the follower agent 3 can converge to the desired position $p_3^*$ at the prespecified time after applying the control algorithm (9.3).

For follower agent 3, the equilibrium points of controller (9.3) satisfies

$$0 = -\left(a + b\frac{\mu'}{\mu}\right)\left(Q_{g_{31}^*}(-\hat{g}_{31}) + Q_{g_{32}^*}(-\hat{g}_{32})\right). \tag{9.16}$$

Multiply $(p_3 - p_3{}^*)^{\mathrm{T}}$ on both sides of (9.16), one has

$$\begin{aligned} 0 &= -(p_3 - p_3^*)^{\mathrm{T}}\left(a + b\tfrac{\mu'}{\mu}\right)\left(Q_{g_{31}^*}(-\hat{g}_{31}) + Q_{g_{32}^*}(-\hat{g}_{32})\right) \\ &= -(p_3 - p_3^*)^{\mathrm{T}}\left(a + b\tfrac{\mu'}{\mu}\right)\left(Q_{g_{31}^*}\left(-\hat{R}(\theta_3)\,g_{31}^3\right) + Q_{g_{32}^*}\left(-\hat{R}(\theta_3)\,g_{32}^3\right)\right), \end{aligned} \tag{9.17}$$

where $p_3$ is the real-time position information of follower agent 3.

According to the convergence analysis of $R\left(\hat{\theta}_3\right)$, one can obtain that $\hat{g}_{31} \to g_{31}$, $\hat{g}_{32} \to g_{32}$ when $t \to T$, that is $\hat{R}(\theta_3) \to R(\theta_3)$. Therefore, (9.17) can be re-described as

$$\begin{aligned} 0 &= -(p_3 - p_3^*)^{\mathrm{T}}\left(a + b\tfrac{\mu'}{\mu}\right)\left(Q_{g_{31}^*}(-g_{31}) + Q_{g_{32}^*}(-g_{32})\right) \\ &= -(p_3 - p_3^*)^{\mathrm{T}}\left(a + b\tfrac{\mu'}{\mu}\right)\left(Q_{g_{31}^*}\tfrac{(p_3 - p_1)}{\|p_3 - p_1\|} + Q_{g_{32}^*}\tfrac{(p_3 - p_2)}{\|p_3 - p_2\|}\right) \\ &= -(p_3 - p_3^*)^{\mathrm{T}}\left(a + b\tfrac{\mu'}{\mu}\right)\left(Q_{g_{31}^*}\tfrac{(p_3 - p_1 - p_3^* + p_3^*)}{\|p_3 - p_1\|} + Q_{g_{32}^*}\tfrac{(p_3 - p_2 - p_3^* + p_3^*)}{\|p_3 - p_2\|}\right), \end{aligned} \tag{9.18}$$

where $-g_{31} = g_{13}$, $-g_{32} = g_{23}$.

Since agents 1 and 2 are leaders and their position are static invariant, the initial position, the real-time position and the desired position are the same, i.e., $p_1^0 = p_1 = p_1^*$, $p_2^0 = p_2 = p_2^*$. Then, (9.18) becomes

$$0 = -(p_3 - p_3^*)^{\mathrm{T}}\left(a + b\tfrac{\mu'}{\mu}\right)\left(Q_{g_{31}^*}\frac{((p_3 - p_3^*) + (p_3^* - p_1^*))}{\|p_3 - p_1^*\|} + Q_{g_{32}^*}\frac{((p_3 - p_3^*) + (p_3^* - p_2^*))}{\|p_3 - p_2^*\|}\right). \tag{9.19}$$

According to the properties of orthogonal projection matrix, it can be obtained that $Q_{g_{31}^*}(p_3^* - p_1^*) = 0$, $Q_{g_{32}^*}(p_3^* - p_2^*) = 0$, then one has

$$0 = -(p_3 - p_3^*)^{\mathrm{T}}\left(a + b\tfrac{\mu'}{\mu}\right)\left(Q_{g_{31}^*}\frac{(p_3 - p_3^*)}{\|p_3 - p_1^*\|} + Q_{g_{32}^*}\frac{(p_3 - p_3^*)}{\|p_3 - p_2^*\|}\right). \tag{9.20}$$

Equation (9.20) holds if and only if $p_3 = p_3^*$. Therefore, $p_3 = p_3^*$ is the unique equilibrium point of (9.3).

Design a Lyapunov function as follows

$$V_3 = \frac{1}{2}\left\| p_3 - p_3^* \right\|^2. \tag{9.21}$$

Take the derivation of $V_3$ and based on the analysis from (9.16) to (9.20), one has

$$\begin{aligned}(V_3)' &= -\left(p_3 - p_3^*\right)^{\mathrm{T}}\left(a + b\tfrac{\mu'}{\mu}\right)\left(Q_{g_{31}^*}\tfrac{(p_3-p_3^*)}{\|p_3-p_1^*\|} + Q_{g_{32}^*}\tfrac{(p_3-p_3^*)}{\|p_3-p_2^*\|}\right)\\ &\leq -\tfrac{2}{\|p_3-p_1^*\|}\left(a + b\tfrac{\mu'}{\mu}\right)\lambda_{Q_{g_{31}^*}} V_3 - \tfrac{2}{\|p_3-p_2^*\|}\left(a + b\tfrac{\dot{\mu}}{\mu}\right)\lambda_{Q_{g_{32}^*}} V_3,\end{aligned} \tag{9.22}$$

where $\lambda_{Q_{g_{31}^*}}$ and $\lambda_{Q_{g_{32}^*}}$ the minimum eigenvalue of matrices $Q_{g_{31}^*}$ and $Q_{g_{32}^*}$, respectively.

Based on (9.22), one can obtain that $(V_3)' = -\alpha_3 V_3 - \beta_3 \frac{\mu'}{\mu} V_3$, where $\alpha_3 = \left(\frac{2}{\|p_3-p_1^*\|}\lambda_{Q_{g_{31}^*}} + \frac{2}{\|p_3-p_2^*\|}\lambda_{Q_{g_{32}^*}}\right)a$, $\beta_3 = \left(\frac{2}{\|p_3-p_1^*\|}\lambda_{Q_{g_{31}^*}} + \frac{2}{\|p_3-p_2^*\|}\lambda_{Q_{g_{32}^*}}\right)b$. According to Lemma 1.1, one has $V_3(t)\begin{cases}\leq \mu^{-\beta_3} e^{-\alpha_3 t} V_3(0), & t \in [0, T)\\ \equiv 0, & t \in [T, \infty)\end{cases}$. Then, one can conclude that $p_3$ can move to $p_3^*$ at the prespecified time $2T$.

**Remark 9.2** For follower agent 3, the convergence of the estimated initial position $\hat{p}_3^0$, as well as the real-time position $p_3$ both depend on the convergence of $\hat{R}(\theta_3)$. In other words, $\hat{p}_3^0 \to p_3^0$ and $p_3 \to p_3^*$ can be guaranteed when $\hat{R}(\theta_3) \to R(\theta_3)$. Meanwhile, it is important to point out that the two convergence processes are synchronized and do not affect each other, and both are able to converge at the specified time $2T$.

The proof of Part I is completed.

**Part II**. The convergence of the estimated rotation matrix $\hat{R}(\theta_i)$, the estimated initial position $\hat{p}_i^0$ and the actual global position $p_i$.

At first, we will prove the convergence of the estimated rotation matrix $\hat{R}(\theta_i)$. For follower agent $i$, the equilibrium points of (9.1) satisfies

$$0 = -\left(a + b\frac{\mu'}{\mu}\right)\left(Q_{\tilde{g}_{jk}}\hat{R}(\theta_i)\, g_{jk}^i\right). \tag{9.23}$$

Multiply $\left(\hat{g}_{jk} - g_{jk}\right)^{\mathrm{T}}$ on both sides of (9.23), yields

$$\begin{aligned}0 &= -\left(a + b\tfrac{\mu'}{\mu}\right)\left(\left(\hat{g}_{jk} - g_{jk}\right)^{\mathrm{T}} Q_{\tilde{g}_{jk}}\hat{R}(\theta_i)\, g_{jk}^i\right)\\ &= -\left(a + b\tfrac{\mu'}{\mu}\right)\left(\left(\hat{R}(\theta_i)\, g_{jk}^i - g_{jk}\right)^{\mathrm{T}} Q_{\tilde{g}_{jk}}\left(\hat{R}(\theta_i)\, g_{jk}^i - g_{jk} + g_{jk}\right)\right).\end{aligned} \tag{9.24}$$

Based on the convergence analysis of Part I, one has $\lim_{t \to 2(i-3)T} \tilde{g}_{jk} = g_{jk}$. Since $Q_{g_{jk}} g_{jk} = 0$, one can obtain that

$$0 = -\left(a + b\frac{\mu'}{\mu}\right)\left(\left(\hat{R}(\theta_i)\, g^i_{jk} - g_{jk}\right)^{\mathrm{T}} Q_{g_{jk}} \left(\hat{R}(\theta_i)\, g^i_{jk} - g_{jk}\right)\right). \tag{9.25}$$

Equation (9.25) holds if and only if $\hat{R}(\theta_i)\, b^i_{jk} - b_{jk} = 0$. Then, one can conclude that $\hat{b}_{jk} = b_{jk}$ is the unique equilibrium point of Eq. (9.25), i.e., $\hat{R}(\theta_i) = R(\theta_i)$.

Select a Lyapunov function as follows

$$V_4 = \frac{1}{2}\left\|\hat{g}_{jk} - g_{jk}\right\|^2. \tag{9.26}$$

Differentiate the function $V_4$, and according to the analysis from (9.23) to (9.25), one has

$$\begin{aligned}(V_4)' &= -\left(a + b\frac{\mu'}{\mu}\right)\left(\left(\hat{R}(\theta_i)\, g^i_{jk} - g_{jk}\right)^{\mathrm{T}} Q_{g_{jk}} \left(\hat{R}(\theta_i)\, g^i_{jk} - g_{jk}\right)\right)\\ &\le -2\left(a + b\frac{\mu'}{\mu}\right)\lambda_{Q_{g_{jk}}} V_4,\end{aligned} \tag{9.27}$$

where $\lambda_{Q_{g_{jk}}} \ge 0$ is minimum eigenvalue of the matrix $Q_{g_{jk}}$.

Inequality (9.27) can be reformulated as

$$(V_4)' \le -\alpha_4 V_4 - \beta_4 \frac{\dot{\mu}}{\mu} V_4, \tag{9.28}$$

where $\alpha_4 = 2a\lambda_{Q_{g_{jk}}}$, $\beta_a = 2b\lambda_{Q_{g_{jk}}}$.

According to Lemma 1.1, one has $V_4(t) \begin{cases} \le \mu^{-\beta_4} e^{-\alpha_4 t} V(0), & t \in [0, T) \\ \equiv 0, & t \in [T, \infty) \end{cases}$. Then, one can draw the conclusion that $\hat{g}_{jk}$ can converge to $g_{jk}$ at the specified time, that is $\hat{R}(\theta_i) \to R(\theta_i)$. The upper limit of specified convergence time is $2(i-3)T + T$ in theory.

Next, we will prove the convergence of the estimated initial position $\hat{p}^0_i$.

For follower agent $i$, the equilibrium points of position estimator (9.2) satisfies

$$0 = -\left(a + b\frac{\mu'}{\mu}\right)\left(Q_{\hat{g}_{ij}}\left(\hat{p}^0_i - \hat{p}^0_j\right) + Q_{\hat{g}_{ik}}\left(\hat{p}^0_i - \hat{p}^0_k\right)\right). \tag{9.29}$$

Multiply $\left(\hat{p}^0_i - p^0_i\right)^{\mathrm{T}}$ on both sides of (9.29), yields

$$
\begin{aligned}
0 &= -\left(a+b\frac{\mu'}{\mu}\right)\left(\hat{p}_i^0-p_i^0\right)^{\mathrm{T}}\left(Q_{\hat{g}_{ij}}\left(\hat{p}_i^0-\hat{p}_j^0\right)+Q_{\hat{g}_{ik}}\left(\hat{p}_i^0-\hat{p}_k^0\right)\right)\\
&= -\left(a+b\frac{\mu'}{\mu}\right)\left(\hat{p}_i^0-p_i^0\right)^{\mathrm{T}}\times\\
&\left(Q_{\hat{g}_{ij}}\left(\hat{p}_i^0-p_i^0+p_i^0-\hat{p}_j^0\right)+Q_{\hat{g}_{ik}}\left(\hat{p}_i^0-p_i^0+p_i^0-\hat{p}_k^0\right)\right).
\end{aligned}
\tag{9.30}
$$

Based on the analysis in Part I, one can obtain that $\hat{p}_j^0 \to p_j^0$, $\hat{p}_k^0 \to p_k^0$ when $t \to 2(i-3)T$, and $\hat{g}_{ij} \to g_{ij}$, $\hat{g}_{ik} \to g_{ik}$ when $t \to 2(i-3)T+T$. Therefore, Eq. (9.30) can be reformulated as

$$
0 = -\left(a+b\frac{\mu'}{\mu}\right)\left(\hat{p}_i^0-p_i^0\right)^{\mathrm{T}}\left(Q_{g_{ij}}+Q_{g_{ik}}\right)\left(\hat{p}_i^0-p_i^0\right). \tag{9.31}
$$

Equation (9.31) holds if and only if $\hat{p}_i^0 - p_i^0 = 0$, so there is a unique equilibrium $\hat{p}_i^0 = p_i^0$.

Design the Lyapunov function as follows

$$
V_5 = \frac{1}{2}\left\|\hat{p}_i^0 - p_i^0\right\|^2. \tag{9.32}
$$

Derive the function $V_5$,

$$
\begin{aligned}
(V_5)' &= -\left(a+b\frac{\mu'}{\mu}\right)\left(\hat{p}_i^0-p_i^0\right)^{\mathrm{T}}\left(Q_{g_{ij}}+Q_{g_{ik}}\right)\left(\hat{p}_i^0-p_i^0\right)\\
&\le -2\left(a+b\frac{\mu'}{\mu}\right)\lambda_{Q_{g_{ij}}+Q_{g_{ik}}}V_5,
\end{aligned}
\tag{9.33}
$$

where $\lambda_{Q_{g_{ij}}+Q_{g_{ik}}} \ge 0$ is the minimum eigenvalues of matrix $(Q_{g_{ij}} + Q_{g_{ik}})$.

Based on (9.33), one has $(V_5)' \le -\alpha_5 V_5 - \beta_5 \frac{\mu'}{\mu} V_5$, where $\alpha_5 = 2a\lambda_{Q_{g_{ij}}+Q_{g_{ik}}}$, $\beta_5 = 2b\lambda_{Q_{g_{ij}}+Q_{g_{ik}}}$.

According to Lemma 1.1, one can obtain that $V_5(t) \begin{cases} \le \mu^{-\beta_5} e^{-\alpha_5 t} V_5(0), & t \in [0, T) \\ \equiv 0, & t \in [T, \infty) \end{cases}$. Then, one can conclude that $\hat{p}_i^0$ can converge to $p_i^0$ at the prescribed time. Theoretically, the upper bound on the convergence time for a specified time is $2(i-2)T$.

Finally, we will prove that the agent $i$, driven by the controller (9.3), converges to the desired position $p_i^*$ at the specified time.

For follower agent $i$, the equilibrium points of controller (9.3) satisfies

$$
0 = -\left(a+b\frac{\mu'}{\mu}\right)\left(Q_{g_{ij}^*}\left(-\hat{g}_{ij}\right)+Q_{g_{ik}^*}\left(-\hat{g}_{ik}\right)\right). \tag{9.34}
$$

Multiply $(p_i - p_i{}^*)^{\mathrm{T}}$ on both sides of (9.34), one has

$$
\begin{aligned}
0 &= -\left(p_i - p_i^*\right)^{\mathrm{T}} \left(a + b\tfrac{\mu'}{\mu}\right) \left(Q_{g_{ij}^*}\left(-\hat{g}_{ij}\right) + Q_{g_{ik}^*}\left(-\hat{g}_{ik}\right)\right) \\
&= -\left(p_i - p_i^*\right)^{\mathrm{T}} \left(a + b\tfrac{\mu'}{\mu}\right) \times \\
&\quad \left(Q_{g_{ij}^*}\left(-\hat{R}\left(\theta_i\right) g_{ij}^i\right) + Q_{g_{ik}^*}\left(-\hat{R}\left(\theta_i\right) g_{ik}^i\right)\right).
\end{aligned} \tag{9.35}
$$

According to the convergence analysis of the estimated rotation matrix in the first part, it can be shown recursively that $\hat{g}_{ij} \to g_{ij}$, $\hat{g}_{ik} \to g_{ik}$ when $t \to 2(i-3)T$, i.e., $\hat{R}(\theta_i) \to R(\theta_i)$. Therefore, (9.35) can be reformulated as

$$
\begin{aligned}
0 &= -\left(p_i - p_i^*\right)^{\mathrm{T}} \left(a + b\tfrac{\mu'}{\mu}\right) \left(Q_{g_{ij}^*}\left(-g_{ij}\right) + Q_{g_{ik}^*}\left(-g_{ik}\right)\right) \\
&= -\left(p_i - p_i^*\right)^{\mathrm{T}} \left(a + b\tfrac{\mu'}{\mu}\right) \left(Q_{g_{ij}^*}\frac{\left(p_i - p_j\right)}{\left\|p_i - p_j\right\|} + Q_{g_{ik}^*}\frac{\left(p_i - p_k\right)}{\left\|p_i - p_k\right\|}\right) \\
&= -\left(p_i - p_i^*\right)^{\mathrm{T}} \left(a + b\tfrac{\mu'}{\mu}\right) \times \\
&\quad \left(Q_{g_{ij}^*}\frac{\left(p_i - p_j - p_i^* + p_i^*\right)}{\left\|p_i - p_j\right\|} + Q_{g_{ik}^*}\frac{\left(p_i - p_k - p_i^* + p_i^*\right)}{\left\|p_i - p_k\right\|}\right),
\end{aligned} \tag{9.36}
$$

where $-g_{ij} = g_{ji}$, $-g_{ik} = g_{ki}$.

Based on the analysis of the estimated initial position in the first part, one can obtain that $p_j \to p_j^*$, $p_k \to p_k^*$ when $t \to 2(i-3)T$. Therefore, (9.36) becomes

$$
\begin{aligned}
0 &= -\left(p_i - p_i^*\right)^{\mathrm{T}} \left(a + b\tfrac{\mu'}{\mu}\right) \times \\
&\quad \left(Q_{g_{ij}^*}\frac{\left(\left(p_i - p_i^*\right) + \left(p_i^* - p_j^*\right)\right)}{\left\|p_i - p_j^*\right\|} + Q_{g_{ik}^*}\frac{\left(\left(p_i - p_i^*\right) + \left(p_i^* - p_k^*\right)\right)}{\left\|p_i - p_k^*\right\|}\right).
\end{aligned} \tag{9.37}
$$

According to the properties of orthogonal projection matrix, it can be obtained that $Q_{g_{ij}^*}\left(p_i^* - p_j^*\right) = 0$, $Q_{g_{ik}^*}\left(p_i^* - p_k^*\right) = 0$, then, one has

$$
0 = -\left(p_i - p_i^*\right)^{\mathrm{T}} \left(a + b\tfrac{\mu'}{\mu}\right) \left(Q_{g_{ij}^*}\frac{\left(p_i - p_i^*\right)}{\left\|p_i - p_j^*\right\|} + Q_{g_{ik}^*}\frac{\left(p_i - p_i^*\right)}{\left\|p_i - p_k^*\right\|}\right). \tag{9.38}
$$

Equation (9.38) holds if and only if $p_i = p_i^*$, so there is a unique equilibrium $p_i = p_i^*$ for controller (9.3).

Design the Lyapunov function as follows

$$
V_6 = \frac{1}{2}\left\|p_i - p_i^*\right\|^2. \tag{9.39}
$$

Derive the function $V_6$, based on the analysis from (9.34) to (9.38), one can obtain that

$$\begin{aligned}(V_6)' &= -(p_i - p_i^*)^{\mathrm{T}} \left(a + b\frac{\mu'}{\mu}\right) \left( Q_{g_{ij}^*} \frac{(p_i - p_i^*)}{\left\| p_i - p_j^* \right\|} + Q_{g_{ik}^*} \frac{(p_i - p_i^*)}{\| p_i - p_k^* \|} \right) \\ &\le -\frac{2}{\left\| p_i - p_j^* \right\|} \left(a + b\frac{\mu'}{\mu}\right) \lambda_{Q_{g_{ij}^*}} V_6 - \frac{2}{\| p_i - p_k^* \|} \left(a + b\frac{\mu'}{\mu}\right) \lambda_{Q_{g_{ik}^*}} V_6,\end{aligned} \tag{9.40}$$

where $\lambda_{Q_{g_{ij}^*}}$ and $\lambda_{Q_{g_{ik}^*}}$ are the minimum eigenvalue of matrices $Q_{g_{ij}^*}$ and $Q_{g_{ik}^*}$, respectively.

According to inequality (9.40), one has $(V_6)' \le -\alpha_6 V_6 - \beta_6 \frac{\mu'}{\mu} V_6$, where $\alpha_6 = \left( \frac{2}{\left\| p_i - p_j^* \right\|} \lambda_{Q_{g_{ij}^*}} + \frac{2}{\| p_i - p_k^* \|} \lambda_{Q_{g_{ik}^*}} \right) a$, $\beta_6 = \left( \frac{2}{\left\| p_i - p_j^* \right\|} \lambda_{Q_{g_{ij}^*}} + \frac{2}{\| p_i - p_k^* \|} \lambda_{Q_{g_{ik}^*}} \right) b$.

According to Lemma 1.1, one can obtain that $V_6(t) \begin{cases} \le \mu^{-\beta_6} e^{-\alpha_6 t} V_6(0), & t \in [0, T) \\ \equiv 0, & t \in [T, \infty) \end{cases}$. Then, one can conclude that $p_i$ can converge to $p_i^*$ at the prescribed time $2(i-2)T$.

The proof of Theorem 9.1 is completed.

## 9.4 Local-Bearing-Only Moving Target Formation Tracking Control

In Section 9.3, the problem of localization estimation and control of stationary target formation is considered. This section generalizes the stationary target localization estimation and formation control integration scheme to track the moving target formation.

### *9.4.1 Moving Target Formation*

The orientation estimation algorithm and the initial position estimation algorithm are the same as in Section 9.3. The leader agents move at a nonzero constant velocity, on the basis of (9.3), a new prescribed-time distributed formation control law is proposed

$$\begin{aligned}(p_i)' = -\left(a + b\frac{\mu'}{\mu}\right) \Big( & Q_{g_{ij}^*} (-\hat{g}_{ij}) + \\ & \left(Q_{g_{ij}^*}(-\hat{g}_{ij})\right)' + Q_{g_{ik}^*} (-\hat{g}_{ik}) + \left(-Q_{g_{ik}^*} \hat{g}_{ik}\right)' \Big).\end{aligned} \tag{9.41}$$

### *9.4.2 Convergence Analysis*

The convergence of the integrated prescribed-time distributed localization estimation and formation control scheme is shown below.

**Theorem 9.2** *Suppose the MASs satisfies Assumption 9.1, applying the integrated prescribed-time distributed localization estimation and formation control scheme (9.1)–(9.2) and (9.41), one can ensure that the estimated rotation matrix $\hat{R}(\theta_i)$ converges to the real rotation matrix $R(\theta_i)$, the estimated initial global position $\hat{p}_i^0$ converges to the global absolute position $p_i^0$ and the follower agents move to the desired position $p_i^*(t)$.*

***Proof*** The convergence analysis of the estimated rotation matrix and the estimated initial position are omitted in this section. The remaining proof is divided into two main parts as follows: in the first part, it will be shown that $p_3$ can move to the desired position at the prescribed time. In the second part, it will be shown that $p_i$ converges to the desired position at the prescribed time.

**Part I**. The convergence of $p_3$.

For follower agent 3, the controller (9.41) satisfies the following equation at the equilibrium point,

$$\begin{aligned} 0 = &-\left(a + b\frac{\mu'}{\mu}\right)\left(Q_{g_{31}^*}\left(-\hat{g}_{31}\right) + \left(Q_{g_{31}^*}\left(-\hat{g}_{31}\right)\right)'\right. \\ &\left.+Q_{g_{32}^*}\left(-\hat{g}_{32}\right) + \left(Q_{g_{32}^*}\left(-\hat{g}_{32}\right)\right)'\right). \end{aligned} \tag{9.42}$$

Multiply $(p_3 - p_3{}^*)^{\mathrm{T}}$ on both sides of (9.42), one has

$$\begin{aligned} 0 = &-\left(p_3 - p_3^*\right)^{\mathrm{T}}\left(a + b\frac{\mu'}{\mu}\right) \times \\ &\left(Q_{g_{31}^*}\left(-\hat{R}\left(\theta_3\right)g_{31}^3\right) + \left(Q_{g_{31}^*}\left(-\hat{R}\left(\theta_3\right)g_{31}^3\right)\right)'\right. \\ &\left.+Q_{g_{32}^*}\left(-\hat{R}\left(\theta_3\right)g_{32}^3\right) + \left(Q_{g_{32}^*}\left(-\hat{R}\left(\theta_3\right)g_{32}^3\right)\right)'\right). \end{aligned} \tag{9.43}$$

According to the convergence analysis of $\hat{R}(\theta_3)$, one can obtain that $\hat{g}_{31} \to g_{31}$, $\hat{g}_{32} \to g_{32}$ when $t \to T$, that is $\hat{R}(\theta_3) \to R(\theta_3)$. Equation (9.43) is equivalent to

$$\begin{aligned} 0 = &-\left(p_3 - p_3^*\right)^{\mathrm{T}}\left(a + b\frac{\mu'}{\mu}\right) \times \\ &\left(Q_{g_{31}^*}\left(-g_{31}\right) + \left(Q_{g_{31}^*}\left(-g_{31}\right)\right)'\right. \\ &\left.+Q_{g_{32}^*}\left(-g_{32}\right) + \left(Q_{g_{32}^*}\left(-g_{32}\right)\right)'\right) \\ = &-\left(p_3 - p_3^*\right)^{\mathrm{T}}\left(a + b\frac{\mu'}{\mu}\right) \times \\ &\left(Q_{g_{31}^*}\frac{(p_3-p_1)}{\|p_3-p_1\|} + \left(Q_{g_{31}^*}\frac{(p_3-p_1)}{\|p_3-p_1\|}\right)'\right. \\ &\left.+Q_{g_{32}^*}\frac{(p_3-p_2)}{\|p_3-p_2\|} + \left(Q_{g_{32}^*}\frac{(p_3-p_2)}{\|p_3-p_2\|}\right)'\right) \\ = &-\left(p_3 - p_3^*\right)^{\mathrm{T}}\left(a + b\frac{\mu'}{\mu}\right) \times \\ &\left(Q_{g_{31}^*}\frac{(p_3-p_1-p_3^*+p_3^*)}{\|p_3-p_1\|} + \left(Q_{g_{31}^*}\frac{(p_3-p_1-p_3^*+p_3^*)}{\|p_3-p_1\|}\right)'\right. \\ &\left.+Q_{g_{32}^*}\frac{(p_3-p_2-p_3^*+p_3^*)}{\|p_3-p_2\|} + \left(Q_{g_{32}^*}\frac{(p_3-p_2-p_3^*+p_3^*)}{\|p_3-p_2\|}\right)'\right), \end{aligned} \tag{9.44}$$

where $-g_{31} = g_{13}$ and $-g_{32} = g_{23}$.

Since agents 1 and 2 are leaders and their position are static invariant, the real-time position is equivalent to the desired position, i.e., $p_1 = p_1^*$, $p_2 = p_2^*$. Meanwhile, it follows from the properties of orthogonal projection matrix that $Q_{g_{31}^*}\left(p_3^* - p_1^*\right) = 0$, $Q_{g_{32}^*}\left(p_3^* - p_2^*\right) = 0$. Therefore, (9.44) becomes

$$\begin{aligned} 0 = &-(p_3 - p_3^*)^{\mathrm{T}}\left(a + b\tfrac{\mu'}{\mu}\right) \times \\ &\left(Q_{g_{31}^*}\tfrac{(p_3-p_3^*)}{\|p_3-p_1^*\|} + \left(Q_{g_{31}^*}\tfrac{(p_3-p_3^*)}{\|p_3-p_1^*\|}\right)' \right. \\ &\left. +Q_{g_{32}^*}\tfrac{(p_3-p_3^*)}{\|p_3-p_2^*\|} + \left(Q_{g_{32}^*}\tfrac{(p_3-p_3^*)}{\|p_3-p_2^*\|}\right)'\right). \end{aligned} \tag{9.45}$$

Equation (9.45) holds if and only if $p_3 - p_3^* = 0$, so there exists a unique equilibrium point $p_3 = p_3^*$ for controller (9.41).

Design the Lyapunov function as follows

$$V_7 = \frac{1}{2}\left\|p_3 - p_3^*\right\|^2. \tag{9.46}$$

Taking the derivative of function $V_7$ yields

$$\begin{aligned} (V_7)' = &-(p_3 - p_3^*)^{\mathrm{T}}\left(a + b\tfrac{\mu'}{\mu}\right)\left(Q_{g_{31}^*}\left(\tfrac{(p_3-p_3^*)}{\|p_3-p_1^*\|} + \tfrac{(p_3-p_3^*)'}{\|p_3-p_1^*\|}\right)\right. \\ &\left.+Q_{g_{32}^*}\left(\tfrac{(p_3-p_3^*)}{\|p_3-p_2^*\|} + \tfrac{(p_3-p_3^*)'}{\|p_3-p_2^*\|}\right)\right) \\ \le &-\tfrac{2}{\|p_3-p_1^*\|}\left(a + b\tfrac{\mu'}{\mu}\right)\lambda_{Q_{g_{31}^*}}V_7 - \tfrac{1}{\|p_3-p_1^*\|}\left(a + b\tfrac{\mu'}{\mu}\right)\lambda_{Q_{g_{31}^*}}(V_7)' \\ &-\tfrac{2}{\|p_3-p_2^*\|}\left(a + b\tfrac{\mu'}{\mu}\right)\lambda_{Q_{g_{32}^*}}V_7 - \tfrac{1}{\|p_3-p_2^*\|}\left(a + b\tfrac{\mu'}{\mu}\right)\lambda_{Q_{g_{32}^*}}(V_7)'. \end{aligned} \tag{9.47}$$

Solving (9.47) yields

$$\begin{aligned} &\left(1 + \left(a + b\tfrac{\mu'}{\mu}\right)\left(\tfrac{\lambda_{Q_{g_{31}^*}}}{\|p_3-p_1^*\|} + \tfrac{\lambda_{Q_{g_{32}^*}}}{\|p_3-p_2^*\|}\right)\right)(V_7)' \\ &\le -\left(\tfrac{2}{\|p_3-p_1^*\|}\lambda_{Q_{g_{31}^*}} + \tfrac{2}{\|p_3-p_2^*\|}\lambda_{Q_{g_{32}^*}}\right)aV_7 \\ &\quad -\left(\tfrac{2}{\|p_3-p_1^*\|}\lambda_{Q_{g_{31}}} + \tfrac{2}{\|p_3-p_2^*\|}\lambda_{Q_{g_{32}^*}}\right)b\tfrac{\mu'}{\mu}V_7. \end{aligned} \tag{9.48}$$

Then inequality (9.48) becomes

$$\begin{aligned} (V_7)' \le &-\frac{\frac{2}{\|p_3-p_1^*\|}\lambda_{Q_{g_{31}^*}} + \frac{2}{\|p_3-p_2^*\|}\lambda_{Q_{g_{32}^*}}}{1+\left(a+b\frac{\mu'}{\mu}\right)\left(\frac{\lambda_{Q_{g_{31}^*}}}{\|p_3-p_1^*\|} + \frac{\lambda_{Q_{g_{32}^*}}}{\|p_3-p_2^*\|}\right)}aV_7 \\ &-\frac{\frac{2}{\|p_3-p_1^*\|}\lambda_{Q_{g_{31}^*}} + \frac{2}{\|p_3-p_2^*\|}\lambda_{Q_{g_{32}^*}}}{1+\left(a+b\frac{\mu'}{\mu}\right)\left(\frac{\lambda_{Q_{g_{31}^*}}}{\|p_3-p_1^*\|} + \frac{\lambda_{Q_{g_{32}^*}}}{\|p_3-p_2^*\|}\right)}b\frac{\mu'}{\mu}V_7. \end{aligned} \tag{9.49}$$

From (9.49), one has $(V_7)' \le -\alpha_7 V_7 - \beta_7 \frac{\mu'}{\mu} V_7$,

$$\alpha_7 = \frac{\frac{2}{\|p_3-p_1^*\|}\lambda_{Q_{g_{31}^*}} + \frac{2}{\|p_3-p_2^*\|}\lambda_{Q_{g_{32}^*}}}{1+\left(a+b\frac{\mu'}{\mu}\right)\left(\frac{\lambda_{Q_{g_{31}^*}}}{\|p_3-p_1^*\|}+\frac{\lambda_{Q_{g_{32}^*}}}{\|p_3-p_2^*\|}\right)} a,$$

$$\beta_7 = \frac{\frac{2}{\|p_3-p_1^*\|}\lambda_{Q_{g_{31}^*}} + \frac{2}{\|p_3-p_2^*\|}\lambda_{Q_{g_{32}^*}}}{1+\left(a+b\frac{\mu'}{\mu}\right)\left(\frac{\lambda_{Q_{g_{31}^*}}}{\|p_3-p_1^*\|}+\frac{\lambda_{Q_{g_{32}^*}}}{\|p_3-p_2^*\|}\right)} b.$$

According to Lemma 1.1, one can obtain that $V_7(t)\begin{cases} \le \mu^{-\beta_7} e^{-\alpha_7 t} V_7(0), & t \in [0, T) \\ \equiv 0, & t \in [T, \infty) \end{cases}$. Then, one can conclude that $p_3$ can converge to $p_3^*$ at the prescribed time $2T$.

**Part II**. The convergence of $p_i$.

For follower agent $i$, the controller (9.41) satisfies the following equation at the equilibrium point

$$\begin{aligned} 0 = &-\left(a+b\frac{\mu'}{\mu}\right)\left(Q_{g_{ij}^*}(-\hat{g}_{ij}) + \left(Q_{g_{ij}^*}(-\hat{g}_{ij})\right)' \right. \\ &\left. +Q_{g_{ik}^*}(-\hat{g}_{ik}) + \left(Q_{g_{ik}^*}(-\hat{g}_{ik})\right)'\right). \end{aligned} \tag{9.50}$$

Multiply $(p_i - p_i{}^*)^{\mathrm{T}}$ on both sides of (9.50),

$$\begin{aligned} 0 = &-(p_i - p_i^*)^{\mathrm{T}}\left(a+b\frac{\mu'}{\mu}\right)\times \\ &\left(Q_{g_{ij}^*}\left(-\hat{R}(\theta_i)\, g_{ij}^i\right) + \left(Q_{g_{ij}^*}\left(-\hat{R}(\theta_i)\, g_{ij}^i\right)\right)' \right. \\ &\left. +Q_{g_{ik}^*}\left(-\hat{R}(\theta_i)\, g_{ik}^i\right) + \left(Q_{g_{ik}^*}\left(-\hat{R}(\theta_i)\, g_{ik}^i\right)\right)'\right). \end{aligned} \tag{9.51}$$

According to the convergence analysis of $\hat{R}(\theta_3)$ in Sect. 9.3, one has $\hat{g}_{ij} \to g_{ij}$, $\hat{g}_{ik} \to g_{ik}$ when $t \to 2(i-3)T + T$, i.e., $\hat{R}(\theta_i) \to R(\theta_i)$. Similarly, one can obtain that $p_j \to p_j^*$, $p_k \to p_k^*$ when $t \to 2(i-3)T$. Therefore, Eq. (9.51) is equivalent to

$$\begin{aligned} 0 = &-(p_i - p_i^*)^{\mathrm{T}}\left(a+b\frac{\mu'}{\mu}\right)\left(Q_{g_{ij}^*}(-g_{ij}) + \right. \\ &\left.\left(Q_{g_{ij}^*}(-g_{ij})\right)' + Q_{g_{ik}^*}(-g_{ik}) + \left(Q_{g_{ik}^*}(-g_{ik})\right)'\right) \\ = &-(p_i - p_i^*)^{\mathrm{T}}\left(a+b\frac{\mu'}{\mu}\right)\left(Q_{g_{ij}^*}\frac{(p_i-p_j^*)}{\|p_i-p_j^*\|} + \right. \\ &\left.\left(Q_{g_{ij}^*}\frac{(p_i-p_j^*)}{\|p_i-p_j^*\|}\right)' + Q_{g_{ik}^*}\frac{(p_i-p_k^*)}{\|p_i-p_k^*\|} + \left(Q_{g_{ik}^*}\frac{(p_i-p_k^*)}{\|p_i-p_k^*\|}\right)'\right) \\ = &-(p_i - p_i^*)^{\mathrm{T}}\left(a+b\frac{\mu'}{\mu}\right)\times \\ &\left(Q_{g_{ij}^*}\frac{(p_i-p_j^*-p_i^*+p_i^*)}{\|p_i-p_j^*\|} + \left(Q_{g_{ij}^*}\frac{(p_i-p_j^*-p_i^*+p_i^*)}{\|p_i-p_j^*\|}\right)' \right. \\ &\left. +Q_{g_{ik}^*}\frac{(p_i-p_k^*-p_i^*+p_i^*)}{\|p_i-p_k^*\|} + \left(Q_{g_{ik}^*}\frac{(p_i-p_k^*-p_i^*+p_i^*)}{\|p_i-p_k^*\|}\right)'\right), \end{aligned} \tag{9.52}$$

where $-g_{ij} = g_{ji}$, $-g_{ik} = g_{ki}$.

From the properties of orthogonal projection matrix, one has $Q_{g_{ij}^*}\left(p_i^* - p_j^*\right) = 0$, $Q_{g_{ik}^*}\left(p_i^* - p_k^*\right) = 0$. Then, Eq. (9.52) becomes

$$
\begin{aligned}
0 = -\left(p_i - p_i^*\right)^{\mathrm{T}}\left(a + b\frac{\mu'}{\mu}\right)\Bigg(Q_{g_{ij}^*}\frac{\left(p_i - p_i^*\right)}{\left\|p_i - p_j^*\right\|} + \\
\left(Q_{g_{ij}^*}\frac{\left(p_i - p_i^*\right)}{\left\|p_i - p_j^*\right\|}\right)' + Q_{g_{ik}^*}\frac{\left(p_i - p_i^*\right)}{\left\|p_i - p_k^*\right\|} + \left(Q_{g_{ik}^*}\frac{\left(p_i - p_i^*\right)}{\left\|p_i - p_k^*\right\|}\right)'\Bigg).
\end{aligned}
\tag{9.53}
$$

Equation (9.53) holds if and only if $p_i - p_i^* = 0$, so there exists a unique equilibrium point $p_i = p_i^*$ for controller (9.41).

Select the Lyapunov function as follows

$$
V_8 = \frac{1}{2}\left\|p_i - p_i^*\right\|^2. \tag{9.54}
$$

Taking the derivative of function $V_8$ yields

$$
\begin{aligned}
(V_8)' = -\left(p_i - p_i^*\right)^{\mathrm{T}}\left(a + b\frac{\mu'}{\mu}\right)\Bigg(Q_{g_{ij}^*}\frac{\left(p_i - p_i^*\right)}{\left\|p_i - p_j^*\right\|} + \\
\left(Q_{g_{ij}^*}\frac{\left(p_i - p_i^*\right)}{\left\|p_i - p_j^*\right\|}\right)' + Q_{g_{ik}^*}\frac{\left(p_i - p_i^*\right)}{\left\|p_i - p_k^*\right\|} + \left(Q_{g_{ik}^*}\frac{\left(p_i - p_i^*\right)}{\left\|p_i - p_k^*\right\|}\right)'\Bigg) \\
\leq -\frac{2}{\left\|p_i - p_j^*\right\|}\left(a + b\frac{\mu'}{\mu}\right)\lambda_{Q_{g_{ij}^*}} V_8 \\
-\frac{1}{\left\|p_i - p_j^*\right\|}\left(a + b\frac{\mu'}{\mu}\right)\lambda_{Q_{g_{ij}^*}} (V_8)' \\
-\frac{2}{\left\|p_i - p_k^*\right\|}\left(a + b\frac{\mu'}{\mu}\right)\lambda_{Q_{g_{ik}^*}} V_8 \\
-\frac{1}{\left\|p_i - p_k^*\right\|}\left(a + b\frac{\mu'}{\mu}\right)\lambda_{Q_{g_{ik}^*}} (V_8)'.
\end{aligned}
\tag{9.55}
$$

By solving (9.55) one can obtain that

$$
\begin{aligned}
(V_8)' \leq -\frac{\frac{2}{\left\|p_i - p_j^*\right\|}\lambda_{Q_{g_{ij}^*}} + \frac{2}{\left\|p_i - p_k^*\right\|}\lambda_{Q_{g_{ik}^*}}}{1 + \left(a + b\frac{\mu'}{\mu}\right)\left(\frac{\lambda_{Q_{g_{ij}^*}}}{\left\|p_i - p_j^*\right\|} + \frac{\lambda_{Q_{g_{ik}^*}}}{\left\|p_i - p_k^*\right\|}\right)} a V_8 \\
-\frac{\frac{2}{\left\|p_i - p_j^*\right\|}\lambda_{Q_{g_{ij}^*}} + \frac{2}{\left\|p_i - p_k^*\right\|}\lambda_{Q_{g_{ik}^*}}}{1 + \left(a + b\frac{\mu'}{\mu}\right)\left(\frac{\lambda_{Q_{g_{ij}^*}}}{\left\|p_i - p_j^*\right\|} + \frac{\lambda_{Q_{g_{ik}^*}}}{\left\|p_i - p_k^*\right\|}\right)} b\frac{\mu'}{\mu} V_8.
\end{aligned}
\tag{9.56}
$$

From (9.56), one has $(V_8)' \leq -\alpha_8 V_8 - \beta_8 \frac{\mu'}{\mu} V_8$, where

$$
\alpha_8 = \frac{\frac{2}{\left\|p_i - p_j^*\right\|}\lambda_{Q_{g_{ij}^*}} + \frac{2}{\left\|p_i - p_k^*\right\|}\lambda_{Q_{g_{ik}^*}}}{1 + \left(a + b\frac{\mu'}{\mu}\right)\left(\frac{\lambda_{Q_{g_{ij}^*}}}{\left\|p_i - p_j^*\right\|} + \frac{\lambda_{Q_{g_{ik}^*}}}{\left\|p_i - p_k^*\right\|}\right)} a,
$$

$$\beta_8 = \frac{\frac{2}{\left\| p_i - p_j^* \right\|} \lambda_{Q_{g_{ij}^*}} + \frac{2}{\left\| p_i - p_k^* \right\|} \lambda_{Q_{g_{ik}^*}}}{1 + \left(a + b\frac{\mu'}{\mu}\right)\left(\frac{\lambda_{Q_{g_{ij}^*}}}{\left\| p_i - p_j^* \right\|} + \frac{\lambda_{Q_{g_{ik}^*}}}{\left\| p_i - p_k^* \right\|}\right)} b.$$

According to Lemma 1.1, one can obtain that $V_8(t) \begin{cases} \leq \mu^{-\beta_8} e^{-\alpha_8 t} V_8(0), & t \in [0, T) \\ \equiv 0, & t \in [T, \infty) \end{cases}$. Then, it can be concluded that $p_i$ can move to $p_i^*$ at the prescribed time $2(i-2)T$.

The proof of the Theorem 9.2 is completed.

## 9.5 Simulation and Experiment

In this section, two numerical simulation examples are provided, and the physical verification is carried out on the unmanned aerial vehicle platform in our laboratory, which used to demonstrate the effectiveness and realizability of the proposed integrated scheme of prescribed time distributed localization and formation control.

### *9.5.1 Numerical Simulation*

In this subsection, the stationary target localization and formation control algorithms (9.1)–(9.3) and the moving target localization and formation control algorithms (9.1) and (9.2), (9.41) are simulated and verified respectively. In the simulation, the orientations of all follower agents are given randomly and the orientation of the leader agent is consistent with the global reference frame. The communication topology of the muti-agent system satisfies the Henneberg construction. The local bearing information is the only data available to the agent.

Two cases are considered, the simulation results are given as follows.

**Case 1** Stationary Target Localization and Formation Control

In this case, a 10-agents multi-agent system is considered, where two agents are leaders and the remaining agents are followers. The desired static coordinate positions are set as: $[-10, 0]$, $[10, 0]$, $[0, 20]$, $[0, -20]$, $[-20, 20]$, $[20, 20]$, $[-20, -20]$, $[20, -20]$, $[-40, 0]$, $[40, 0]$. The parameters for the orientation estimation and the initial position estimation algorithms are selected as follows: $h_1 = 10$, $a_1 = 0.45$, $b_1 = 3.5$, and the upper limit of the prescribed convergence time is set to $T_1 = 5$, the unit is seconds. The parameters of the controller are selected as follows: $h_2 = 20$, $a_2 = 1$, $b_2 = 2$, and the upper limit of the prescribed convergence time is set to $T_2 = 15$. The initial estimated values of the orientation and the initial position are given arbitrarily, the initial value of the control input is set to zero. Applying the distributed localization and formation algorithms (9.1)–(9.3), and the simulation results are shown in Figs. 9.3, 9.4, 9.5, 9.6 and 9.7.

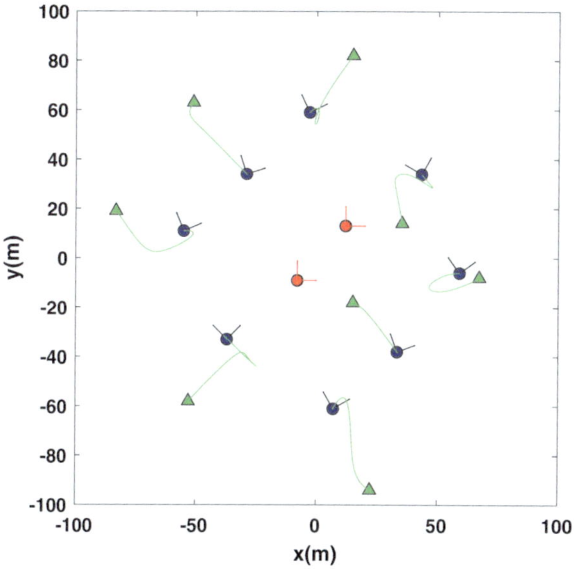


**Fig. 9.3** The convergence trajectories of the estimated initial position

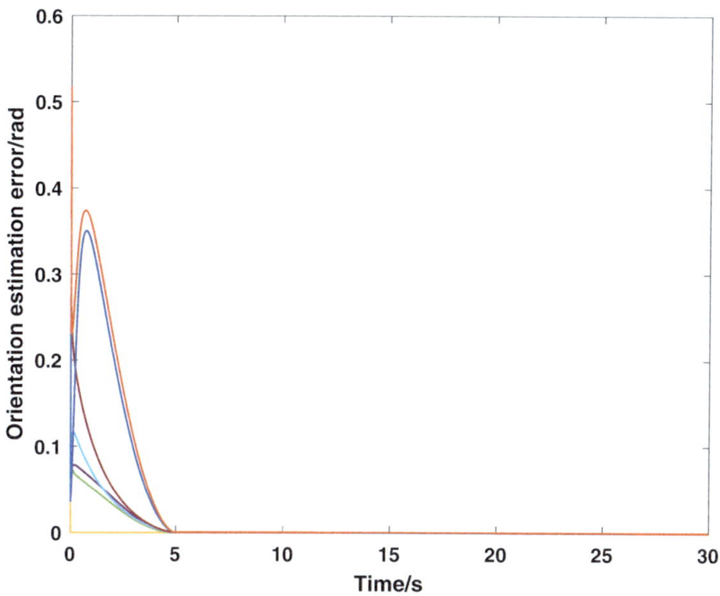


**Fig. 9.4** The orientation estimation error curves

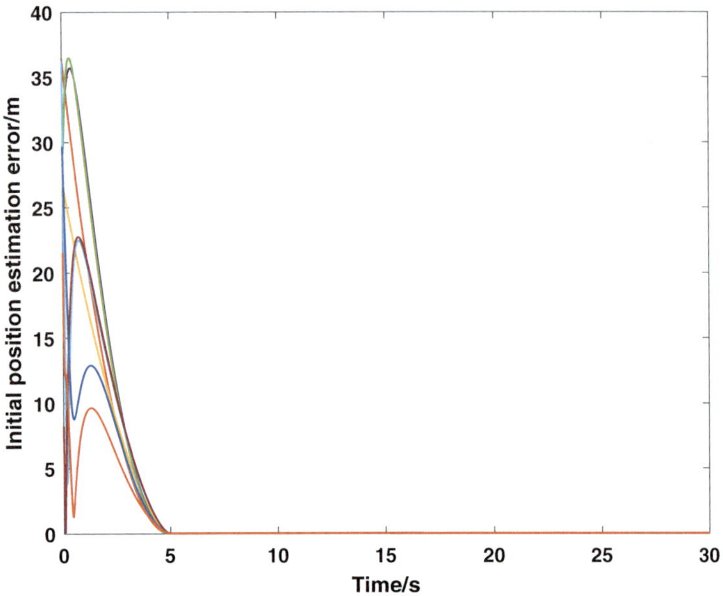


**Fig. 9.5** The initial position estimation error curves

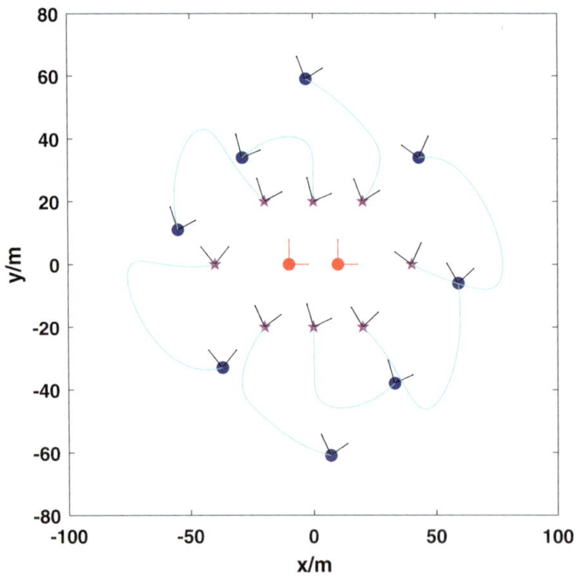


**Fig. 9.6** The stationary target formation process trajectories

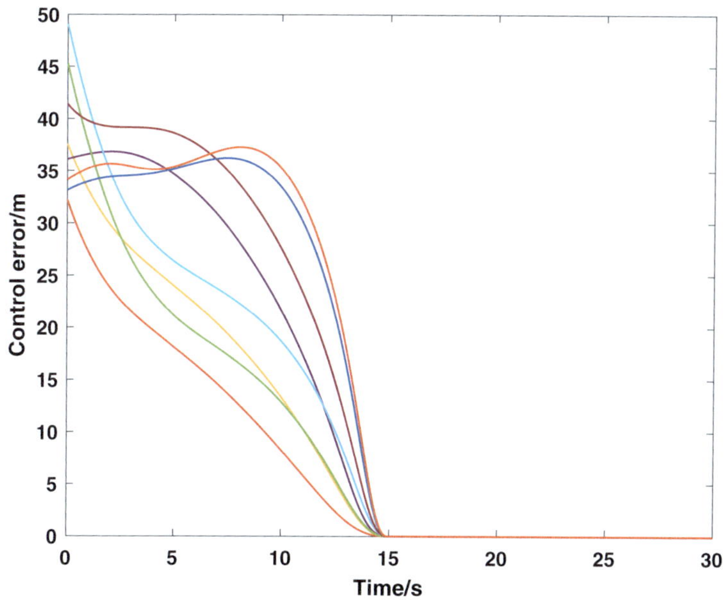


**Fig. 9.7** The control error curves

The convergence trajectories of the follower agents are shown in Fig. 9.3, where the green triangles are the initial estimated coordinates and the green curves are the convergence trajectories of the estimated coordinates. Figure 9.4 shows the orientation estimation error curves of the follower agents, the estimation error is defined as $\left\| \hat{R}(\theta_i) - R(\theta_i) \right\|$, it can be seen that the estimation error can converge to zero at time $T = 5$. The initial position estimation error curves are shown in Fig. 9.5, one can find that the estimation error $\| \hat{p}_i^0 - p_i^0 \|$ can converge to zero at the fifth second.

The formation process of the multi-agent system is shown in Fig.9.6, where the blue circles represent the initial positions of the agents, the colored curves represent the moving trajectories of the agents after applying the controller (9.3), and the purple pentagrams represent the desired positions of the agents (i.e., stationary targets). Figure 9.7 shows the control error curves of the follower agents. It can be observed that the control error $\| p_i - p_i^* \|$ converges to zero at the 15th second.

**Case 2** Moving Target Localization and Formation Control

In this case, a 4-agents multi-agent system is considered, where two agents are leaders and the remaining agents are followers. The trajectory equation of leader agent 1 is $\begin{cases} x_1(t) = 15 + 0.6 * t, \\ y_1(t) = 5 + 0.2 * t. \end{cases}$ The trajectory equation of leader agent 2 is $\begin{cases} x_2(t) = 15 + 0.6 * t, \\ y_2(t) = -5 + 0.2 * t. \end{cases}$ The desired bearing of the follower agent 3 is

$g_{31}^* = \begin{bmatrix} 1 \\ 0 \end{bmatrix}$, $g_{32}^* = \begin{bmatrix} 0.707 \\ -0.707 \end{bmatrix}$. The desired bearing of the follower agent 4 is $g_{41}^* = \begin{bmatrix} 0.707 \\ 0.707 \end{bmatrix}$, $g_{42}^* = \begin{bmatrix} 1 \\ 0 \end{bmatrix}$.

The parameters for the orientation estimation and the initial position estimation algorithms are selected as follows: $h_1 = 20$, $a_1 = 0.45$, $b_1 = 4.5$, and the upper limit of the prescribed convergence time is set to $T_1 = 10$. The parameters of the controller are selected as follows: $h_2 = 120$, $a_2 = 1$, $b_2 = 25$, and the upper limit of the prescribed convergence time is set to $T_2 = 20$. The initial estimated values of the orientation and the initial position are given arbitrarily, the initial value of the control input is set to zero. Applying the distributed localization and formation algorithms (9.1), (9.2) and (9.41), and the simulation results are shown in Figs. 9.8, 9.9 and 9.10.

The convergence trajectories of the initial estimated position of the follower agents are shown in Fig. 9.8, where the green triangles are the initial estimated coordinates and the green curves are the convergence trajectories of the estimated coordinates. Figure 9.9 shows the orientation estimation error curves of the follower agents, it can be seen that the estimation error can converge to zero at time $T = 10$.

The formation process of the multi-agent system is shown in Fig. 9.10, where the purple and green curves are the real movement trajectories of the leader agents 1 and 2, while the red and yellow curves are the movement trajectories of the two follower agents under the action of the controller. It can be found that the two follower agents are ultimately able to achieve the desired moving target formation at the prescribed time.

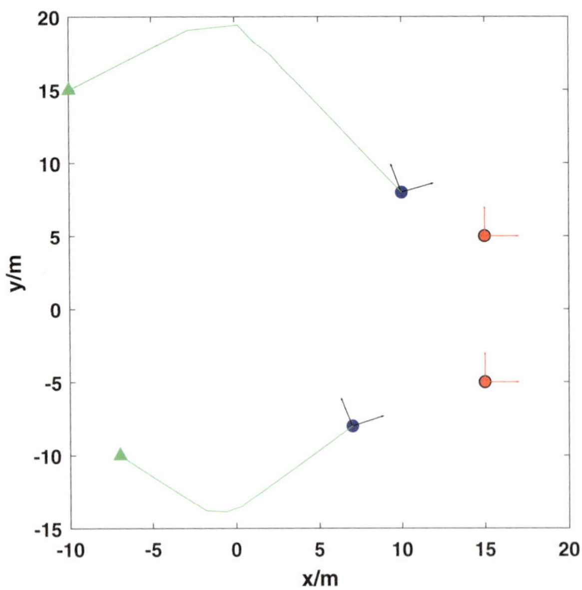


**Fig. 9.8** The convergence trajectories of the estimated initial position

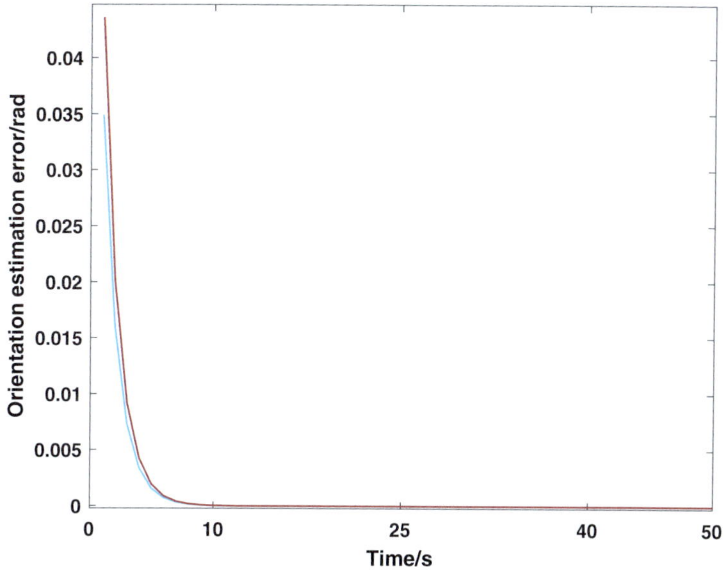


**Fig. 9.9** The orientation estimation error curves

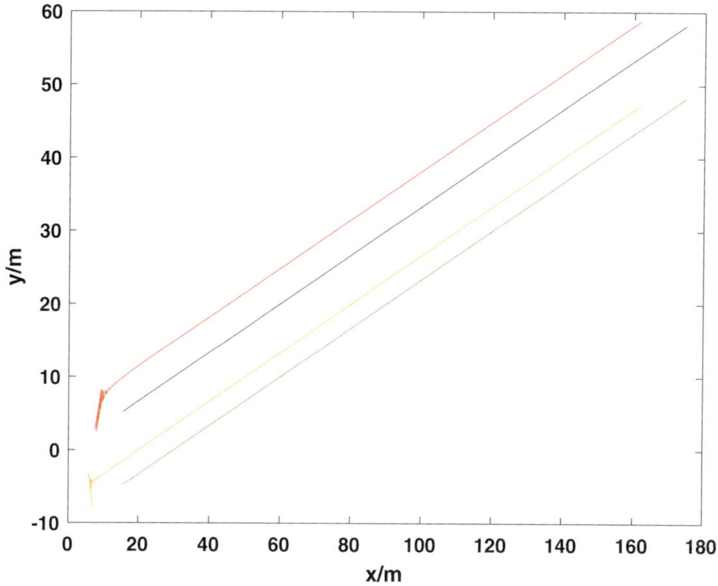


**Fig. 9.10** The moving target formation process trajectories

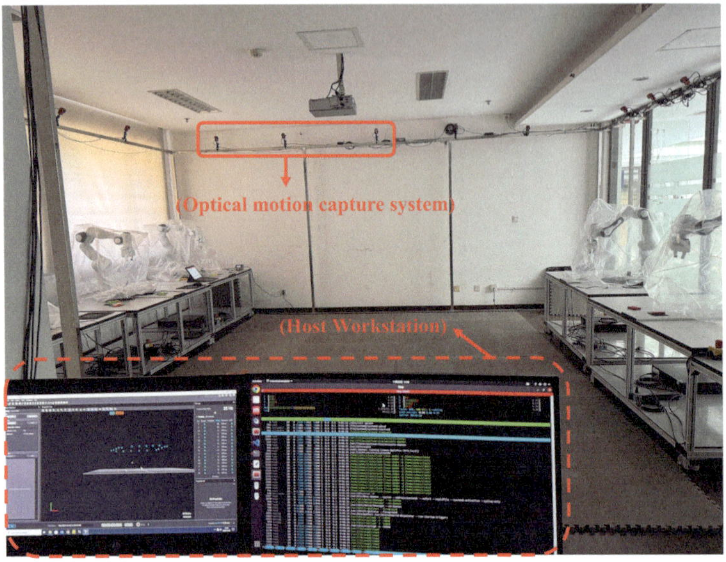


**Fig. 9.11** The high-precision optical motion capture system

### 9.5.2 *Quad-Rotor UAV Experiment*

In this subsection, the effectiveness and practicality of the distributed localization estimation and formation control algorithms (9.1)–(9.3) are verified using the mini quad-rotor UAV, and the experimental results are discussed and analyzed.

OptiTrack system is a high-precision optical motion capture system, as shown in Fig. 9.11. The motion-capture system with 20 cameras is installed in an $6m \times 3m \times 3m$ space. The communication between the motion capture system and the base station is based on the protocol defined by Robot Operating System. In this experiment, ten mini quad-rotor UAV crazyflies with passive spherical markers are used to test the validity of the algorithms (9.1)–(9.3).

The global absolute position of the UAVs satisfy the Henneberg configuration, and the communication topology between the UAVs is shown in Fig. 9.12, where $L$ represents the leader agents and $F$ represents the followers.

The estimate of the initial position of the follower UAV is given randomly. The parameters are chosen as follows: $h = 10$, $a = 0.45$, $b = 3.5$, and the upper bound of the prescribed convergence time is set to $T = 10$. Applying the proposed distributed localization estimation and formation control algorithms (9.1)–(9.3). Figure 9.13 illustrates the formation flight process of the ten crazyflies, and it can be found that the UAVs are able to form the desired formation. The experimental results show that the proposed prescribed-time integrated distributed localization estimation and formation control scheme achieves the desired formation of UAVs.

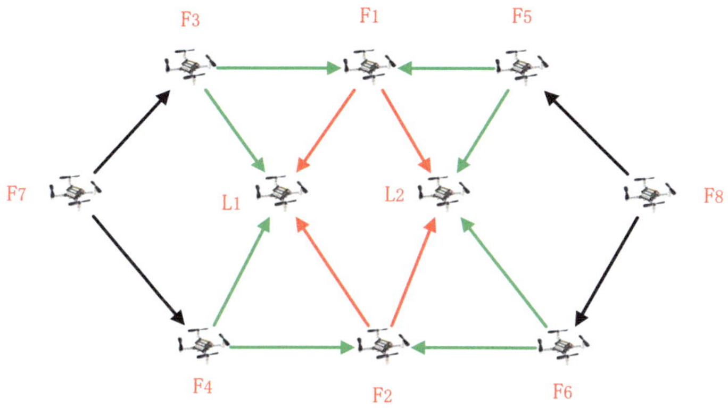


**Fig. 9.12** The communication topology between UAVs

Simulation and experimental results verify the effectiveness and realizability of the proposed prescribed-time integrated distributed localization estimation and formation control scheme. As a result, the following conclusions can be drawn: the proposed distributed localization estimation algorithm based on local bearing information can ensure the precise orientation alignment of the multi-agent system, and realize the stationary target and moving target formation control in the local reference frame.

## 9.6 Conclusions

In this chapter, the proposed local-bearing-based prescribed-time integrated distributed localization and formation scheme solves the localizability and controllability of multi-agent systems under directed graph. The designed prescribed-time orientation estimation algorithm guarantees that the local reference frame of the follower agent is aligned with the global one. The developed two prescribed-time distributed formation control schemes realize the stationary and moving target formation with unknown initial positions. The global convergence is derived based on the cascade system and the mathematical induction method. Both simulation and experiment results are provided to prove the effectiveness of the theoretical results.

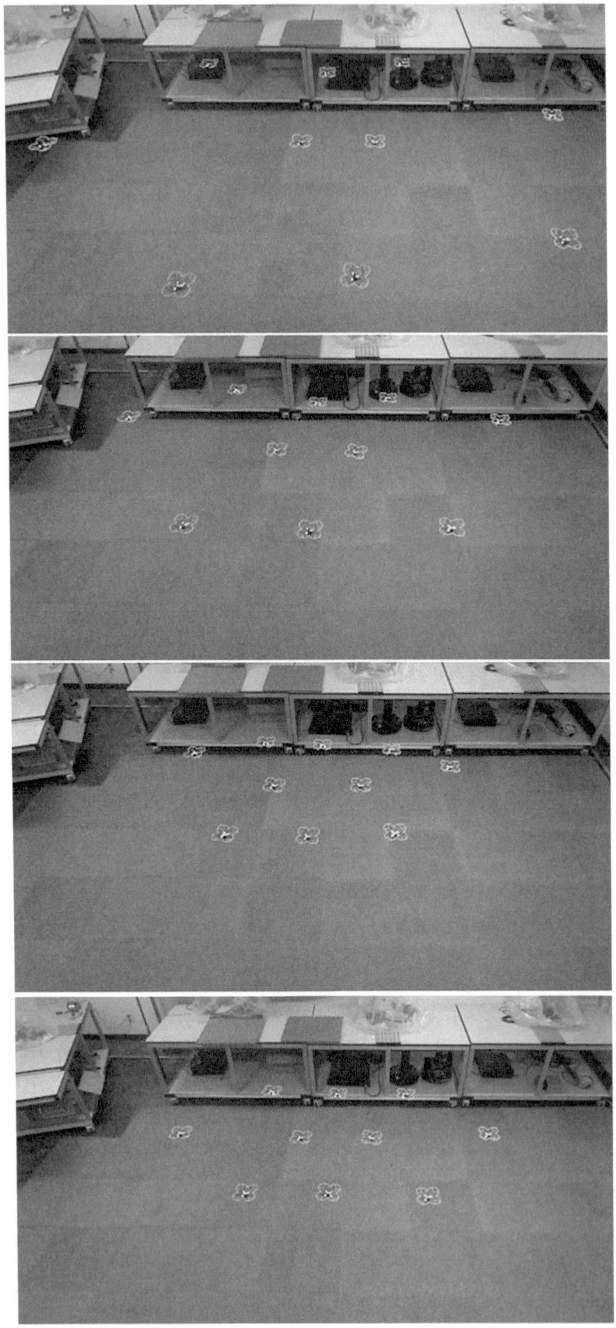

**Fig. 9.13** The formation flight process of the ten crazyflies

## References


1. Ren, W.: Multi-vehicle consensus with a time-varying reference state. Syst. Control Lett. **56**, 474–483 (2007)
2. Sun, Z., Anderson, B.: Rigid formation control with prescribed orientation. In: Proceedings of IEEE International Symposium on Intelligent Control, pp. 639–645. Sydney, Australia (2015)
3. Zelazo, D., Franchi, A., Giordano, P.: Rigidity theory in SE(2) for unscaled relative position estimation using only bearing measurements. In: Proceedings of European Control Conference, pp. 2703–2708. Strasbourg, France (2014)
4. Eren, T.: Formation shape control based on bearing rigidity. Int. J. Control **85**(9), 1361–1379 (2012)
5. Zhao, S., Zelazo, D.: Bearing-based distributed control and estimation in multi-agent systems. In: Proceedings of European Control Conference, pp. 2207–2212. Linz, Austria (2015)
6. Bishop, A., Shames, I., Anderson, B.: Stabilization of rigid formations with direction-only constraints. In: Proceedings of IEEE Conference on Decision and Control and European Control Conference, pp. 746–752. Orlando, USA (2011)
7. Franchi, A., Giordano, P.: Decentralized control of parallel rigid formations with direction constraints and bearing measurements. In: Proceedings of IEEE Conference on Decision and Control, pp. 5310–5317. Maui, USA (2012)
8. Zhao, S., Zelazo, D.: Bearing rigidity and almost global bearing-only formation stabilization. IEEE Trans. Autom. Control **61**(5), 1255–1268 (2016)
9. Zhao, S., Zelazo, D.: Localizability and distributed protocols for bearing-based network localization in arbitrary dimensions. Automatica **69**, 334–341 (2016)
10. Zhao, S., Li, Z., Ding, Z.: Bearing-only formation tracking control of multiagent systems. IEEE Trans. Autom. Control **64**(11), 4541–4554 (2019)
11. Zhao, S., Lin, F., Peng, K., et al.: Finite-time stabilization of cyclic formations using bearing-only measurements. Int. J. Control **87**(4), 715–727 (2014)
12. Trinh, M., Mukherjee, D., Zelazo, D., et al.: Finite-time bearing only formation control. In: Proceedings of IEEE Conference on Decision and Control, pp. 1578–1583. Melbourne, Australia (2017)
13. Tran, Q., Trinh, M.H., Zelazo, D., et al.: Finite time bearing-only formation control via distributed global orientation estimation. IEEE Trans. Control Netw. Syst. **6**(2), 702–712 (2019)
14. Li, Z., Tnunay, H., Zhao, S., et al.: Bearing-only formation control with prespecified convergence time. IEEE Trans. Cybern. **52**(1), 620–629 (2022)
15. Oh, K., Ahn, H.: Formation control and network localization via orientation alignment. IEEE Trans. Autom. Control **59**(2), 540–545 (2014)
16. Trinh, M., Lee, B., Ye, M., et al.: Bearing-based formation control and network localization via global orientation estimation. In: Proceedings of IEEE Conference on Control Technology and Applications, pp. 1084–1089. Copenhagen, Denmark (2018)
17. Li, X., Luo, X., Zhao, S.: Globally convergent distributed network localization using locally measured bearings. IEEE Trans. Control Netw. Syst. **7**(1), 245–253 (2019)